Mini EURO Conference on Improving Healthcare: New Challenges, New Approaches

Journal of Physics: Conference Series Volume 616

Coimbra, Portugal
30 March – 1 April 2015

ISBN: 978-1-5108-0583-5
ISSN: 1742-6588

Printed from e-media with permission by:

Curran Associates, Inc.
57 Morehouse Lane
Red Hook, NY 12571

Some format issues inherent in the e-media version may also appear in this print version.

Printed by Curran Associates, Inc. (2015)

For permission requests, please contact the Institute of Physics
at the address below.

Institute of Physics
Dirac House, Temple Back
Bristol BS1 6BE UK

Phone: 44 1 17 929 7481
Fax: 44 1 17 920 0979

techtracking@iop.org

Additional copies of this publication are available from:

Curran Associates, Inc.
57 Morehouse Lane
Red Hook, NY 12571 USA
Phone: 845-758-0400
Fax: 845-758-2634
Email: curran@proceedings.com
Web: www.proceedings.com

TABLE OF CONTENTS

SIMULATION IN HEALTHCARE

PREFACE

The XXXI *Euro Mini Conference on Improving Healthcare: new challenges, new approaches* took place in Coimbra, Portugal, from the 30th March to the 1st April 2015. This volume contains 17 contributed papers, presented in the Conference.

Recent advances in medicine allow us to live longer and healthier lives. These advances have been made possible through the joint contribution of very different scientific fields. Healthcare is nowadays better understood as a multidisciplinary field, with increasing number of challenges that can be better tackled by joint collaboration of researchers with different scientific backgrounds.

The challenges that Healthcare is facing have not only to do with the ability of providing better services to all (treatments, preventive medicine, better diagnosis tools), with an increased focus on improved personalized treatment options, but also with the need to tackle an increased pressure felt by Healthcare systems: increased number of patients, some of them requiring expensive treatments, with a consequent increase in the workload for health institutions that most of the times have to deal with important budget restrictions.

The Conference provided a forum in which researchers coming from different scientific disciplines discussed and shared their experience regarding methodological approaches to tackle different healthcare challenges. The Conference welcomed contributions from operational research, industrial engineering, medicine, medical physics, management, computational biology, bioinformatics and health economics, among others.

The Conference had 93 enrolled participants, from 22 different nationalities: Australia, Belgium, Brazil, Canada, Hong Kong, Ireland, Israel, Italy, Lebanon, Poland, Portugal, Romania, Serbia, Singapore, South Africa, Spain, Sweden, Switzerland, The Netherlands, United Arab Emirates, United Kingdom and United States of America. These participants had very different academic backgrounds: operations research, medicine, medical physics, engineering, biomedicine, medicine, among others. Moreover, participants had also very different professional backgrounds: researchers from academia, researchers from private companies, clinicians, etc. This diversity contributed to the fulfillment of the Conference objectives. Among these participants there were 20 PhD students presenting their research work, many of them for the first time.

The Conference Program was composed by a total of 70 talks, 4 of which invited talks. Our invited speakers were Alexandre Quintanilha, Ben Heijmen, David Craft and Sally Braislford.

The conference was supported by EURO (the Association of European Operational Research Societies), ORAHS (EURO Working Group on Operational Research Applied to Health Services), APDIO (Portuguese Operational Research Society), Inesc-Coimbra, Faculty of Economics of University of Coimbra, FCT (Portuguese Science Foundation) and COMPETE under project grant PTDC/EIA-CCO/121450/2010.

Peer review statement

All papers published in this volume of *Journal of Physics: Conference Series* have been peer reviewed through processes administered by the proceedings Editors. Reviews were conducted by expert referees to the professional and scientific standards expected of a proceedings journal published by IOP Publishing.

Mini EURO Conference on Improving Healthcare: new challenges, new approaches IOP Publishing
Journal of Physics: Conference Series **616** (2015) 012001 doi:10.1088/1742-6596/616/1/012001

Development and initial testing of a pulse oximetry prototype for measuring dental pulp vitality

Cerqueira M, Ferreira M, Caramelo F[1]
IBILI, Azinhaga de St. Comba, Celas
300-548 COIMBRA

E-mail: fcaramelo@fmed.uc.pt

Abstract. The guiding principle of endodontic treatment is to preserve teeth while maintaining its aesthetic and functional roles. To accomplish this goal the assessment of teeth pulp vitality is very important since it will determine the procedures that should be adopted and define the therapy strategy. Currently, the most commonly tests for determining dental pulp state are the thermal and the electrical tests, which are based on nerve response and, because of that, have a relatively high rate of false positives and false negatives cases. In this work we present a simple test to be used in the clinical setting for evaluating noninvasively the existence of blood perfusion in dental pulp. This test is based on pulse oximetry principle that was devised to indirectly measure the amount of oxygen in blood. Although pulse oximetry has already demonstrated its usefulness in clinical environment its usage for the determination of dental pulp vitality has been frustrated by several factors, notably the absence of a suitable sensor to the complex shape of the various coronary teeth. We developed a suitable sensor and present the first trials with promising results, regarding the ability for distinguish teeth with and without blood perfusion.

1. Introduction

According to the World Health Organization, 90% of world population suffers from oral cavity diseases which, most of the times, are painful [1]. To give responses to these problems better dental materials, new clinical procedures and more accurate diagnosis tests are being created in a daily basis. Despite all advances, the success of the therapy is still highly dependent on the diagnosis. For example, whenever the dentist suspects that the dental pulp may be compromise it is crucial to he knows very well its state in order to adopt the most correct procedures that will ensure the preservation of pulp tissue and avoid inflammation (pulpitis) or necrosis. An accurate diagnosis of the pulp state is essential in many situations, such as to anticipate a procedure for tooth restoration, to follow up and monitoring the pulp after a tooth trauma and as supporting knowledge to many other pathologies [2].
Many of the current methods for diagnosing the pulp state stimulate the A-δ nerve fibers, which gives information about nerve condition but not on vascularization. Nonetheless, there is some correlation between blood perfusion and nervous response because in the absence of perfusion the A-δ fibers enter in anoxia stopping its function. This means that these methods are considered admissible [3, 4] which coupled to the fact of being simple, inexpensive and having a quick answer make them the most frequent tests currently used in the clinical practice.

[1] To whom any correspondence should be addressed.

The sensorial response of the pulp is achieved using a cold, a hot or an electrical stimulus. Cold agents induce the contraction of the fluid within the dentinal tubules, followed by its rapid efflux which then activates the A-δ nerve fibers and causes pain [5]. As cold agents ice sticks and ethyl chloride gas were initially employed being substituted, in recent years, by more effective gases such as dichlorodifluoromethane and carbon dioxide. Both dichlorodifluoromethane and carbon dioxide permit cooler temperatures producing a more intense response of the A-δ nerve fibers. However, repeated application of cold tend to reduce the rate of fluid relocation within the dentinal tubules resulting in a lower pain for a short period of time, turning the cold test not very reliable. As an alternative there are methods that use heat sensation instead, but as with cold agents the pain results from the fluid dynamics at the dentin-pulp complex [5, 6].

The purpose of the electric pulp test is also to stimulate A-δ fibers that remain intact in the pulp-dentin complex. An electrical stimulus is applied in the tooth surface, after being properly insulated, causing an alteration in neuronal ionic membrane that gives rise to an action potential in the Ranvier nodes of the myelinated fibers. Once the voltage reaches the pain threshold the patient feels discomfort in the tooth under examination revealing the integrity of nerve fibers [5, 7, 8].

The sensitive methods are very simple to use however the rate of false positives and false negatives is 10%-16% making these methods unreliable [9,10]. Other more objective methods, based on the state of the dental pulp, have been proposed to overcome this difficulty. Laser doppler flowmetry, photoplethysmography, spectrophotometry and pulse oximetry are the most important of these techniques that try to assess the state of the pulp by evaluating the blood flow in the dental cavity through optical properties [11]. These methods are supported by the principle that the pulp tissue can have an adequate blood supply despite not being necessarily innervated [10]. And they are also based on the assumption that light interact with blood enabling to distinguish the presence and absence of blood perfusion within a region [12]. Among the methods that rely on light interaction pulse oximetry is the less expensive and is well accepted in the clinical setting. Therefore we decide to exploit the same rationale to develop a device for measuring the vitality of dental pulp [9].

2. Pulse oximetry

Light and matter interaction can be modeled by Beer-Lambert equation that links the intensity attenuation with properties of the medium and light wavelength (equation 1) [13]:

$$I = I_0 e^{-\mu(\lambda) x}, \tag{1}$$

where I is the light intensity emerging from a medium of length x and attenuation factor μ when the incidence light has intensity I_0. Notice that attenuation depends on the wavelength of the incidence light. In these conditions the fraction of transmitted light can then be expressed by the exponential term that depends on the wavelength of the incidence light and on the traveling length,

$$f = \frac{I}{I_0} = e^{-\mu(\lambda) x} \tag{2}$$

Since oxygen concentration in blood has a particular impact on light attenuation it is possible to determine the percentage of oxygen saturation in blood as long as different wavelengths are used to determine the same number of transmitted light fractions [14]. Pulse oximetry commonly use wavelengths of 660 nm (red) and 940 nm (infrared – IR). This choice is driven by four important aspects: (i) absorption is mainly in wavelengths lower than 600 nm (blue, green and yellow) due to skin pigmentation; (ii) absorption of near infrared by water; (iii) absorption of visible light in red region is much higher for deoxygenated hemoglobin (Hb) than for oxygenated hemoglobin (HbO2); and (iv) the IR transparency is greater for Hb than for HbO2 [15]. Pulse oximeters also take advantage of blood pulsation during cardiac cycle: in systole arteries contain more blood than in diastole having, in turn, a larger amount of absorbent substances, such as hemoglobin and also a larger optical distance. This means that light attenuation increases during systole. This dynamic portion of total absorption allows to discriminate absorption due to static components (venous blood, a constant amount of

Mini EURO Conference on Improving Healthcare: new challenges, new approaches IOP Publishing
Journal of Physics: Conference Series **616** (2015) 012001 doi:10.1088/1742-6596/616/1/012001

arterial blood and other invariants such as skin pigmentation) and absorption due to the pulsatile component (blood flow in the arteries). These components are also denominated as DC and AC components of the total absorption, respectively. Arranging measures of DC and AC components from the two wavelengths it is possible to have an estimate of oxygen saturation [16]. This measure is known as Ratio of Ratios (RR) and is defined by

$$RR = \frac{\ln(\frac{AC_R}{DC_R})}{\ln(\frac{AC_{IR}}{DC_{IR}})} \quad (3)$$

where AC_R and DC_R are the AC and DC components, respectively, of the red light and in denominator we have the same ratio for the infrared light.

3. Methods

3.1. Sensor design

Pulse oximeters are normally used to take measurements from finger or earlobe, which are locations anatomically favorable for placing sensors. However, teeth present different shapes and sizes making more difficult to devise a form to place the sensor that should be tightly positioning while ensuring parallelism between emitter and receptor. The approach that we adopt was based on forceps as it can be applied to any tooth and can be easily handled. The sensor holder was designed using CAD software (Inventor®) and a plastic model was then built using rapid prototyping (Fig 1).

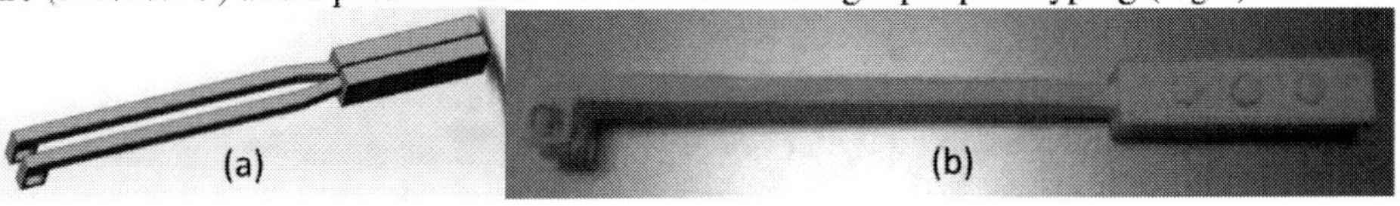

Figure 1. (a) CAD design of the sensor holder; (b) plastic prototype.

At the tip of holder three LEDs were mounted: a red (660 nm), an infrared (950 nm) and also a green LED (540 nm) to account for the enamel and dentin contribution to the light attenuation [17, 18]. At the other tip of the holder, a phototransistor was placed and the signal was acquired using a UD128A8D B&B module connected to a PC. The same electronic module was used to turn on and off the LEDs sequentially. The power needed to maintain the minimum required light intensity was achieved through a power driver (ULN2003A) and signal amplification from the phototransistor was accomplished using a differential amplifier (AD626).

3.2. Sensor testing

Sensor repeatability and linearity was assessed by some simple tests. Repeatability was evaluated by measuring the signal produced for each LED when positioning at two different distances (in contact and 3.0 mm apart) from the phototransistor. Measures were made for each LED individually and mean, standard deviation and coefficient of variation of the acquired signal were calculated as figures of merit to assess response and repeatability. Noise was also addressed by running tests in dark conditions, i.e., acquiring signal without turning on the LEDs.
Linearity was evaluated using a colloidal mixture with four different dilutions fractions. The main idea was to verify Beer-Lambert law by obtaining a linear relationship between the log of the light signal and concentration of the solutions.

3.3. Clinical testing

Preliminary clinical tests were carried out with two reasons: to evaluate the capacity of using the device in the clinical setting and to examine its ability for discriminating vital from non-vital teeth. A dentist conducted 29 tests in four different patients and evaluated pulp vitality of each tooth. Before

Mini EURO Conference on Improving Healthcare: new challenges, new approaches IOP Publishing
Journal of Physics: Conference Series **616** (2015) 012001 doi:10.1088/1742-6596/616/1/012001

measuring the dentist insulated the tooth with a suitable rubber barrier and dried it with an air jet. The measuring was performed for 30s, which has facilitated the accurate positioning of the sensor in the labial and lingual faces of teeth.
Results were analysed through logistic regression method using IBM SPSS v20 statistical platform and assuming a significance level of 5% ($\alpha = 0.05$).

4. Results

4.1. Sensor testing
Table 4.1 shows the results (empiric units) obtained for repeatability tests ran at 0.0 and 3mm distance between LEDs and the phototransistor.

Table 4.1. Light intensity (empiric units) for 0 mm and 3 mm distances between emitter and receiver.

		R	IR	G
0 mm	$\mu \pm \sigma$	413±56	3734±13	2046±67
	C.V.	13.56%	0.35%	3.27%
3 mm	$\mu \pm \sigma$	2588±82	3738±10	2956±73
	C.V.	3.17%	0.27%	2.47%

In dark conditions the mean signal in the different channels (red, infrared and green) is 47, 47 and 46 respectively and the standard deviation is 10, 11 and 8 respectively.
Sensor linearity testing was performed for different concentrations of a colloidal solution. Results of light intensity in function of concentration were fitted with a linear regression after a logarithm transformation of the phototransistor signal. Response linearity was confirmed visually but also by linear regression. The adjusted linear model for the red LED is statistically significant ($F(1,10) = 34{,}296$, $p < 0.001$) indicating a 75% percentage explanation of the variation of the signal (adj. $R^2 = 0.752$) by concentration variation. For the infrared LED the fitted linear model indicates an explanation by the tested concentration of approximately 78% of the variation of the signal (adj. $R^2 = 0.784$) and the model is statistically significant ($F(1,10) = 36.283$, $p < 0.001$). Finally, for the green LED it can be observed that the linear model shows that concentration explain about 41% of variation of light intensity (adj. $R^2 = 0.412$) and the model is also statistically significant ($F(1,10) = 8.720$, $p = 0.014$).

4.2. Clinical testing
The main goal of the work was the classification of a tooth as vital or not vital. Therefore it was very important to extract from the raw signal some numerical indicators that will enable classifying dental vitality. Taking into account the nature of the signal it was intuitive to take advantage of the AC and DC components to compose numerical indicators for the classification. Nonetheless, we decided to use an extra LED that provides further information. We rearranged and simplified all measures as follows:

$$R = \frac{\ln(AC_R/DC_R)}{\ln(AC_{IR}/DC_{IR})} \tag{4}$$

$$R_G = \ln(AC_G/DC_G) \tag{5}$$

where the subscripts in AC and DC components are referred to the red (R), infrared (IR) and green (G) LEDs, respectively.

Although we have collected further measures the number of non-vital teeth in the sample was a limitation for the analysis. For that reason we decided to analyse similar teeth (molar and pre-molar). Table 4.2 shows results of the ratios according to the equations presented above for the selected cases:

Table 4.2. Results for vital and non-vital teeth in the sample.

	Vital						
	No	No	Yes	Yes	Yes	Yes	No
R	0.984	0.785	1.006	0.994	1.005	0.991	0.993
RG	−2.773	-2.738	-3.129	-2.540	-2.986	-2.059	-2.829

A logistic regression model was fitted to the results assuming vitality classification as dependent variable and the ratio scores as predictor variables. The model obtained shows a significant adjustment (Hosmer and Lemshow, $\chi(4) = 0.00$; $p = 1.000$) and the predictor variables explain the dependent variable (Cox and Snell $R^2 = 0.745$, Nagelkerke $R^2 = 1.000$). These excellent values are due to the few cases that were analysed which made simple the adjustment to a predictor model. The accuracy obtained was 100% which can be considered significant if compared to the null model (when the classification is random) that presents 57.1% accuracy rate. The model can be described analytically by the equation:

$$P = \frac{e^{4724.814 - 94.496 \times Rg - 5019.286 \times R}}{1 + e^{4724.814 - 94.496 \times Rg - 5019.286 \times R}} \qquad (6)$$

where P is the probability of having a non-vital tooth.

5. Discussion and conclusion

We presented a new device for measuring dental pulp vitality based on pulse oximetry principles. Results indicate a functional device with good reproducibility, low noise and reasonable linearity properties. However, the construction of the instrument revealed some difficulties that are points of concern. Besides the problem of having a general form of placing the sensor on teeth surfaces the utilization of a parallel design between emitter and receiver poses a mechanical problem that has to be carefully tackled. In our case this can be observed by comparing the values obtained when the sensor was tested at two different distances. Small changes in signal (Table 4.1) can be observed between the two distances for each LED which may be explained by the different solid angle at the receiver. This means that different teeth will produce distinct signals in part because of the size and this aspect has to be taken into account in future developments.

Another aspect that should be notice is a dependence on phototransistor sensibility to the emitter wavelength. It seems that this problem do not have great impact in the final results, nonetheless it can be resolved by implementing a software gain to each LED since they are turning on/off sequentially.

Clinical trials exposed other kind of problems: the design of the instrument has to be improved in order to facilitate its usage and teeth have to be perfectly dry otherwise noise is dramatically increased. Results of clinical tests do not permit to generalize and consider them as conclusive. The sample is small and not randomized, which explain the excellent fitting results that may be regarded as a sign of overfitting. Nevertheless, they show a promising trend because it performs much better than a random approach (just guessing) and using only two features.

To address the validation problem, a larger clinical trial has to be design, enabling to draw definitive conclusions.

6. References

[1] Ferreira M, Albuquerque B, Siri P, Ginjeira A and Capelas J 2007 Glossário de Termos endodônticos *Revista Portuguesa de Estomatologia e Medicina Dentária Cirurgia Maxilofacial* **48** 247-255.

[2] Goodis H 2002 History of Pulp Biology *Seltzer and Bender's Dental Pulp* ed Hargreaves K and Goodis H (Quintessence Publishing) pp 1-11

[3] Abd-Elmeguid A and Yu D 2009 Dental Pulp Neurophysiology: Part 1. Clinical and Diagnostic Implications *Journal of the Canadian Dental Association* **75** 55-59

[4] Garg N and Garg A 2010 Diagnostic Procedures *Textbook of Endodontics* ed Garg N and Garg A (New Dehli: Jaypee Brothers Medical Publishers) pp 64-67

[5] Ford T and Patel S 2004 Technical equipment for assessment of dental pulp status *Endodontic Topics* **7** 2–13

[6] Amish D, Jyoti M and Jitendra 2007 Assessment of Pulp Vitality in Children *Journal of Dental Sciences* **2** 44-46

[7] Karayilmaz H and Kirzioglu Z 2011 Comparison on the reliability of laser Doppler flowmetry, pulse oximetry and electric pulp tester in assessing the pulp vitality of human teeth *Journal of Oral Rehabilitation* **38** 340-347

[8] Gopikrishna V, Tinagupta K and Kandaswamy D 2007 Comparison of Electrical, Thermal, and Pulse Oximetry Methods for Assessing Pulp Vitality in Recently Traumatized Teeth *Journal of Endodontics* **33** 531-535

[9] Dastmaichi N, Jafarzadeh H and Moradi Saeed 2012 Comparision of the Efficacy of a Custom-made Pulse Oximeter Probe with Digital Electrical Pulp Tester, Cold Spray, and Rubber Cup for Assessing Pulp Vitality *Journal of Endodontics* **38** 1182-1186

[10] Abd-Elmeguid A and Yu D 2009 Dental Pulp Neurophysiology: Part 2. Current Diagnostic Tests to Assess Pulp Vitality *Journal of the Canadian Dental* **75** 139-143

[11] Chen E and Abbott V 2009 Dental Pulp Testing: A Review *International Journal of Dentistry* 1-12

[12] Vaghela D and Sinha A 2011 Pulse oximetry and laser Doppler flowmetry for diagnosis of pulpal vitality *Journal of Interdisciplinary Dentistry* **1** 14-21

[13] Wieben O 1997 Light Absorbance Pulse Oximetry *Design of Pulse Oximeters* ed Webster J (New York: Taylor & Francis) pp 40-55

[14] Farmer J 1997 Blood Oxygen Measurement *Design of Pulse Oximeters* ed Webster J (New York: Taylor & Francis) pp 21-39

[15] Wukitsch M, Petterson M, Tobler D and Pologe J 1988 Pulse Oximetry: Analysis of Theory, Technology and Practice *Journal of Clinical Monitoring* **4** 290-301

[16] Abdallah O and Bolz A 2011 Adaptive Filtering by Non-Invasive Vital Signals Monitoring and Diseases Diagnosis *Adaptive Filtering Applications* ed Garcia L (InTech) pp 157-172

[17] Kahan R, Gulabivala K, Snook M and Setchell D 1996 Evaluation of a Pulse Oximeter and Customized Probe for Pulp Vitality Testing *Journal of Endodontics* **22** 105-109

[18] Ruchala J 1997 Applications of Pulse Oximetry *Design of Pulse Oximeters* ed Webster J (New York: Taylor & Francis) pp 214-236

Diabetic peripheral neuropathy assessment through texture based analysis of corneal nerve images

Susana F. Silva[1], Sofia Gouveia[2], Leonor Gomes[2], Luís Negrão[3], Maria João Quadrado[1,4], José Paulo Domingues[1,5] and António Miguel Morgado[1,5]

[1]IBILI - Institute for Biomedical Imaging and Life Sciences, Faculty of Medicine, University of Coimbra, Coimbra, Portugal
[2]Department of Endocrinology, Diabetes and Metabolism, Coimbra Hospital University Center, Coimbra, Portugal
[3]Department of Neurology, Coimbra Hospital University Center, Coimbra, Portugal
[4]Department of Ophthalmology, Coimbra Hospital University Center, Coimbra, Portugal
[5]Department of Physics, Faculty of Sciences and Technology, University of Coimbra, Coimbra, Portugal

Email: Susana.fig.silva@gmail.com

Abstract. Diabetic peripheral neuropathy (DPN) is one common complication of diabetes. Early diagnosis of DPN often fails due to the non-availability of a simple, reliable, non-invasive method. Several published studies show that corneal confocal microscopy (CCM) can identify small nerve fibre damage and quantify the severity of DPN, using nerve morphometric parameters. Here, we used image texture features, extracted from corneal sub-basal nerve plexus images, obtained *in vivo* by CCM, to identify DPN patients, using classification techniques. A SVM classifier using image texture features was used to identify (DPN vs. No DPN) DPN patients. The accuracies were 80.6%, when excluding diabetic patients without neuropathy, and 73.5%, when including diabetic patients without diabetic neuropathy jointly with healthy controls. The results suggest that texture analysis might be used as a complementing technique for DPN diagnosis, without requiring nerve segmentation in CCM images. The results also suggest that this technique has enough sensitivity to detect early disorders in the corneal nerves of diabetic patients.

1. Introduction

The use of morphologic information extracted from corneal confocal microscopy images of the sub-basal nerve plexus has been proposed for diagnosing and staging diabetic peripheral neuropathy (DPN) [1-3]. This research effort is well justified. DPN is one of the more common complications of diabetes, being the main cause of chronic disability in diabetic patients. It is present in about 8% of newly diagnosed patients, affecting up to 50% of the patients after 25 years of disease and is associated to 50-75% of non-traumatic amputations [4, 5]. Early diagnosis, which is very important to define higher risk patients, often fails or occurs only when patients became symptomatic due to the non-availability of a simple, reliable, non-invasive method [6]. The standard methods for DPN diagnosis, which are based on the evaluation of neuropathic symptoms and signs and on electrophysiology measurements, are not adequate for detecting the early changes occurring in small nerve fibres. These can be detected measuring the intra-epidermal nerve fibre density through skin biopsy [7-9]. However this is an invasive, complex procedure. Corneal confocal microscopy (CCM) [10, 11] offers the possibility of inspecting *in vivo* small nerve fibres by non-invasive optical methods, using instrumentation already available at the ophthalmology departments of central hospitals.

It was already shown that diabetic patients have lower nerve density [12], even for short diabetes duration [13, 14], that CCM can accurately report the extent of corneal nerve damage and repair using fibre density and branching measurements [1], and that nerve tortuosity correlates with neuropathy severity. It was also demonstrated that corneal nerve fibre loss correlates with intra-epidermal nerve fibre loss [2], with the parameter corneal nerve fibre length showing higher discriminative capacity to diagnose DPN [15]. Recent works focus on the development of accurate methods for automatic segmentation and analysis of corneal nerves on CCM images [16-21] and on the reproducibility of CCM in the evaluation of corneal nerves morphology, using manual, semi-automated and automatic methods [22-24]. A review on this research area was recently published [25].

Here we present results of a pilot study designed to evaluate the use of texture analysis of corneal nerve images obtained with corneal confocal microscopy for DPN detection. Texture based analysis used image statistics obtained from the Gray Level Co-occurrence Matrix (GLCM) and from the Gray Level Run Length Matrix (GLRLM).

2. Methods

2.1. Study design

The study included 12 diabetic patients (type 2, insulin-treated, with mean age of 58±10 years), followed at the Department of Endocrinology, Diabetes and Metabolism of Coimbra Hospital University Center (CHUC), and 8 age-matched non-diabetic control individuals (mean age: 54±7 years).

The study was conducted according to guidelines of the Declaration of Helsinki and submitted to approval by the Medical Ethics Committee of the Faculty of Medicine of the University of Coimbra. Informed consent was obtained from all participants after explanation of the nature and possible consequences of the study. Clinical examinations followed the international consensus guidelines for diagnosis and management of DPN [26] and the Michigan Neuropathy Screening Instrument (MNSI) [27], comprising two separate assessments: a 15-item questionnaire and a lower extremity examination, that includes feet appearance evaluation, ankle reflex testing and sensory deficit evaluation (superficial pain, touch perception and vibrating sensation). Physical examination and questionnaire evaluation was done by medical doctors, in a random order, without information of other test results. The patients' clinical history was verified to ensure that peripheral neuropathy is a consequence only of diabetes.

All individuals underwent electromyography (EMG) at the Department of Neurology of CHUC, using a Nicolet Biomedical EA4 (Nicolet Biomedical, Madison, WI, USA) recording device. EMG measurements comprised nerve conduction evaluation, motor (peroneal) and sensory (sural) nerve conduction velocity (NCV) and amplitudes, as well as cutaneous sympathetic response.

CCM images were obtained at the Department of Ophthalmology of CHUC, using a Heidelberg Retinal Tomograph equipped with a Cornea Rostock Module (Heidelberg Engineering, Heidelberg, Germany). The 384x384 pixels images correspond to a 400μm x 400μm area and were saved in JPEG format. All individuals underwent bilateral examination by an ophthalmologist. We recorded 275 nerve images from healthy individuals and 356 images from diabetic patients. Image samples are shown in Figure 1.

2.2. DPN evaluation based on image texture features

Images were used as acquired, without any pre-processing. Textural analysis was based on first-order image statistics (variance), second-order statistics based on the Gray Level Co-occurrence Matrix (GLCM) [24] and higher order statistics, based on the Gray Level Run Length Matrix (GLRLM) [28, 29]. The second-order features extracted were energy, entropy, correlation, homogeneity, dissimilarity, cluster shade and cluster prominence. The correlation, energy and homogeneity parameters were obtained for orientations of 0, 45, 90 and 135°. The higher-order features extracted were Short Run Emphasis (SRE), Long Run Emphasis (LRE), Gray-Level Non-uniformity (GLN), Run Length Non-Uniformity (RLN), Run Percentage (RP), Low Gray-Level Run Emphasis (LGRE), High Gray-Level Run Emphasis (HGRE), Short Run Low Gray-Level Emphasis (SRLGE), Short Run High Gray-Level Emphasis (SRHGE), Long Run Low Gray-Level Emphasis (LRLGE) and Long Run High Gray-Level Emphasis (LRHGE). These higher-order features were also calculated for orientations of 0, 45, 90 and 135°. The total number of features extracted from each image was 61. All features were extracted using a Matlab program written for that purpose.

A support vector machine (SVM) system was used for supervised classification of each image in the DPN and No DPN groups and to classify each image of the No DPN group as Mild or Moderate DPN. Training and classification was done using the Weka Data Mining software [30] and Weka LibSVM [31], using the radial basis function kernel. The best features were automatically selected according to a rank given by the square of the weight assigned by a SVM classifier and then transformed by Principal Component Analysis (PCA), performed in the feature space. The SVM classification performance was assessed using both 10-fold cross-validation and leave-one-out validation.

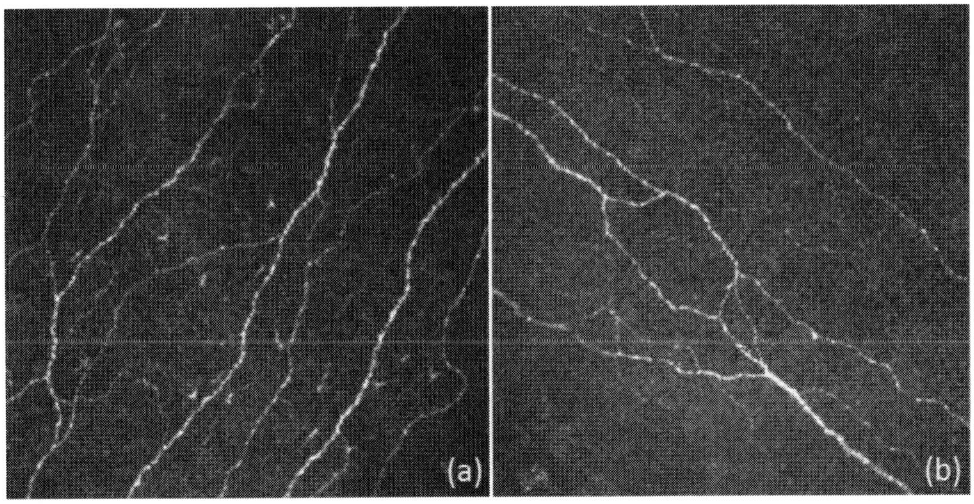

Figure 1. Images of corneal sub-basal nerve plexus obtained by corneal confocal microscopy: (a) healthy cornea; (b) diabetic cornea

3. Results

3.1. Evaluation of study participants

According to their baseline evaluation, the diabetic patients were divided in three groups: absent (4 patients, 53±11 years), mild (5 patients, 58±9 years) and moderate DPN (3 patients, 60±9 years). No patient presented severe DPN. Table 1 summarizes the data of the study participants while Table 2 contains the results of EMG measurements.

Table 1. Summary of the four participants groups (baseline evaluation).

Group (DPN Severity)	Mean Age±Std.Dev. (years)	MNSI Quest. (0-15)	MNSI Scoring (0-10)	EMG diagnosis
Control	54±7	0	0	Normal
Absent	53±11	0 - 3	0 - 1	Normal
Mild	58±9	4 - 6	2 - 3.5	DPN/Normal
Moderate	60±9	6 - 8	2 - 3	DPN

Table 2. Nerve conduction studies results (mean ± std. dev.).

	Motor nerve		Sensory nerve		Cutaneous
Groups	NCV (m/s)	Amplitude (μV)	NCV (m/s)	Amplitude (μV)	Sympathetic response (s)
Control	46.8 ± 2.2	5.1 ± 2.0	47.9 ± 3.5	18.0 ± 6.7	1.98 ± 0.12
No DPN	43.9 ± 2.4	3.5 ± 1.2	43.9 ± 4.4	12.3 ± 5.5	1.86 ± 0.35
Mild DPN	44.1 ± 4.4	4.4 ± 3.0	46.5 ± 5.6	12.9 ± 2.2	2.16 ± 0.35
Moderate DPN	38.3 ± 5.3	2.4 ± 1.0	43.0 ± 6.4	5.2 ± 2.9	1.89 ± 0.01

For motor NCV and sensory nerve conduction amplitudes, significant differences were found between Control and Moderate DPN groups. Although statistical tests have not revealed any significant difference between the existing groups for the sensory NCV and for the sympathetic response, sensory NCV and amplitudes were established by the neurophysiologist as abnormal (below the limit) in patients with moderate degree of DPN.

Normal NCV values were observed for all patients without DPN and for the majority of patients with mild DPN. Nevertheless, the NCV and amplitude values were always lower for patients with mild and moderate DPN, for both motor and sensory nerves.

The results show large-fibre abnormalities in all patients with moderate DPN, and in one patient with mild DPN. Progressive axonal losses, demonstrated by decreased motor amplitudes and reduced sensory amplitudes, were found according to DPN severity. The low values of both motor and sensory NCV indicate that moderate DPN cases occur with severe nerve demyelination.

3.2. *DPN evaluation based on image texture features*

From the set of 61extracted image texture features, 7 were automatically selected for use with the SVM classifier. The SVM was used to establish a supervised classification method to classify each image in the 'with DPN' or 'without DPN' groups. Two different evaluations were done: on the first evaluation the Control individuals and the No DPN diabetic patients were both considered as 'without DPN', while the Mild e Moderate DPN patients were tagged as 'with DPN'. On the second evaluation, the images from No DPN diabetic patients were not included. Table 3 presents the results, obtained for both evaluations, for 10-fold cross-validation and leave-one out validation.

Table 3. SVM classification of images from individuals with and without DPN.

'with DPN' or 'without DPN'	Correctly Classified Instances	
	10-fold cross-validation	Leave-one out validation
Including No DPN diabetic patients (631 images)	73.5 %	73.7 %
Not including No DPN diabetic patients (525 images)	80.6 %	79.8 %

We also evaluated the ability of the classifier to classify correctly the images of the "with DPN" in the groups "mild DPN" and "moderate DPN". The results are presented in Table 4.

Table 4. SVM classification of images from individuals with DPN.

	Correctly Classified Instances	
	10-fold cross-validation	Leave-one out validation
mild DPN' or 'moderate DPN'	79.3 %	78.9 %

In order to evaluate the reproducibility of the classifier, the images included in the 'with DPN' or 'without DPN' groups were randomly divided between two sub-groups. This way, two sets of 'with DPN' and 'without DPN' images were obtained. The first set had 124 images from DPN patients and 138 images from healthy controls. The second set included 126 images from DPN patients and 137 images from healthy controls. The diabetic patients without DPN were not included in this reproducibility assessment. The results are presented in Table 5.

Table 5. Reproducibility assessment of SVM classifier.

'with DPN' or 'without DPN'	Correctly Classified Instances	
	10-fold cross-validation	Leave-one out validation
Set A (262 images)	82.1 %	82.1 %
Set B (263 images)	77.2 %	76.8 %

4. Discussion

An automated image analysis system for objective detection and grading of DPN is essential for establishing CCM as a reference clinical technique for DPN diagnosis and assessment. The methods proposed by different research groups are based on automatic algorithms for nerve segmentation and extraction of morphology parameters. These parameters are then analyzed using statistical between groups comparison tests, for identifying differences between controls and patients groups, or using classification techniques to assign individuals to those groups [16, 17, 20]. This approach requires robust automatic nerve segmentation algorithms able to perform even with lower quality images.

In this work we considered a different methodology, where the nerve images were analysed as a whole, without performing a nerve segmentation step. We choose to use classification techniques based on features obtained from texture analysis of corneal nerve images. Although, to the best of our knowledge, this approach was never used for analysing CCM images, texture analysis has been extensively used for medical image processing, mainly in radiology and nuclear medicine, but also for medical imaging based on optical radiation. Recently, classification techniques based on texture features were used for classifying hysteroscopical images [32]. Hysteroscopy produces images of superficial endometrial vessels, which have structural similarities with the images of the corneal sub-basal nerve plexus obtained by CCM.

We evaluated the use of this methodology to discriminate between individuals with and without DPN. This was done in two different ways: joining the diabetic patients without DPN with the healthy individuals to establish the 'without DPN" group or excluding those patients and using just the healthy individuals and the diabetic patients with DPN in the evaluation. In the first case we obtained an accuracy of 73.5% while in the second evaluation the observed accuracy was 80.6%.

We also used the same methodology to grade the images from patients with DPN in the mild and moderate groups. The achieved accuracy was 79.3%. In all instances the classifier used the same set of 7 attributes selected from the 61 image attributes computed.

The classifier reliability was evaluated by randomly assigning every image of the 'with DPN' and 'without DPN' groups to one of two subsets. This way, we setup two pairs of smaller 'with DPN' and 'without DPN' groups. In this procedure we obtained an average accuracy of 79.7% ± 3.5%. The small standard deviation obtained is a good indicator of the classifier reproducibility.

This are very promising results particularly because they were achieved for a low number of image features and without any image pre-processing or any adaptive process in the extraction of those texture features. It is reasonable to expect higher accuracies with more robust feature extraction methods.

The lower accuracies observed, when using images from diabetic patients diagnosed as not having diabetic neuropathy, suggests that the analysis of corneal nerve images is more sensitive to early nerve disorders in diabetic patients that the standard electrophysiological measurements or clinical examinations and assessments recommended by the international consensus guidelines for diagnosis and management of DPN and included in the Michigan Neuropathy Screening Instrument. It was already shown that morphometric analysis of corneal nerve images has enough sensitivity to detect early disorders in the corneal nerves of diabetic patients [1, 33, 34]. Our results suggest that texture analysis of corneal nerve images have the same potential. However, further studies are required to confirm such potential.

As far as we know, this is the first attempt to use the information contained in CCM images of the corneal sub-basal nerve plexus to assess DPN, without measuring morphology parameters after performing nerve segmentation. The evaluation of DPN based on the morphometric of corneal nerves has the important advantage of being based on parameters with a clear clinical meaning. Corneal nerve morphometric parameters provide direct quantification of small fibres disorders. The Nerve Fibre Length and Nerve Fibre Density parameters are related to enhanced nerve degeneration, while the Nerve Branching Density is associated to nerve regenerative capacity [1]. The methods based on these parameters are able to quantify corneal nerve fibre degeneration and regeneration and can be used assess therapeutic efficacy of treatments. It will be difficult to obtain this amount of information using texture based methods for analysing corneal nerve images. However, as our results suggest, these methods may be suited for screening purposes.

References

[1] Malik RA, Kallinikos P, Abbott CA, van Schie CH, Morgan P, Efron N, et al. *Diabetologia* 2003 May;46(5):683-8.
[2] Quattrini C, Tavakoli M, Jeziorska M, Kallinikos P, Tesfaye S, Finnigan J, et al. *Diabetes* 2007 Aug;56(8):2148-54.
[3] Tavakoli M, Quattrini C, Abbott C, Kallinikos P, Marshall A, Finnigan J, et al. *Diabetes Care* 2010 Aug;33(8):1792-7.
[4] Gooch C, Podwall D. *Neurologist* 2004 Nov;10(6):311-22.
[5] Vinik AI, Park TS, Stansberry KB, Pittenger GL. *Diabetologia* 2000 Aug;43(8):957-73.
[6] Rahman M, Griffin SJ, Rathmann W, Wareham NJ. *Diabet Med* 2003 May;20(5):368-74.

[7] Smith AG, Howard JR, Kroll R, Ramachandran P, Hauer P, Singleton JR, et al. *Journal of the Neurological Sciences* 2005 Jan 15;228(1):65-9.
[8] Lauria G, Hsieh ST, Johansson O, Kennedy WR, Leger JM, Mellgren SI, et al. *European Journal of Neurology* 2010 Jul;17(7):903-E49.
[9] Malik RA, Veves A, Walker D, Siddique I, Lye RH, Schady W, et al. *Acta Neuropathol* 2001 Apr;101(4):367-74.
[10] Patel DV, McGhee CN. *Clin Experiment Ophthalmol* 2007 Jan;35(1):71-88.
[11] Erie JC, McLaren JW, Patel SV. *Am J Ophthalmol* 2009 Nov;148(5):639-46.
[12] Rosenberg ME, Tervo TM, Immonen IJ, Muller LJ, Gronhagen-Riska C, Vesaluoma MH. *Invest Ophthalmol Vis Sci* 2000 Sep;41(10):2915-21.
[13] Popper M, Quadrado MJ, Morgado AM, Murta JN, Van Best JA, Muller LJ. *Invest Ophthalmol Vis Sci* 2005;46(Suppl S):2194.
[14] Midena E, Brugin E, Ghirlando A, Sommavilla M, Avogaro A. *J Refract Surg* 2006 Nov;22(9 Suppl):S1047-S1052.
[15] Ahmed A, Bril V, Orszag A, Paulson J, Yeung E, Ngo M, et al. *Diabetes Care* 2012 Apr;35(4):821-8.
[16] Scarpa F, Grisan E, Ruggeri A. *Invest Ophthalmol Vis Sci* 2008 Nov;49(11):4801-7.
[17] Dabbah MA, Graham J, Petropoulos IN, Tavakoli M, Malik RA. *Med Image Anal* 2011 Oct;15(5):738-47.
[18] Petropoulos IN, Alam U, Fadavi H, Marshall A, Asghar O, Dabbah MA, et al. *Invest Ophthalmol Vis Sci* 2014 Apr;55(4):2071-8.
[19] Ferreira A, Morgado AM, Silva JS. *Computer Methods and Programs in Biomedicine* 2012 Jul;107(1):53-60.
[20] Holmes TJ, Pellegrini M, Miller C, Epplin-Zapf T, Larkin S, Luccarelli S, et al. *Invest Ophthalmol Vis Sci* 2010 Sep;51(9]:4480-91.
[21] Scarpa F, Zheng X, Ohashi Y, Ruggeri A. *Invest Ophthalmol Vis Sci* 2011 Aug;52(9]:6404-8.
[22] Petropoulos IN, Manzoor T, Morgan P, Fadavi H, Asghar O, Alam U, et al. *Cornea* 2013 May;32(5]:E83-E89.
[23] Dehghani C, Pritchard N, Edwards K, Russell AW, Malik RA, Efron N. *Cornea* 2014 Jul;33(7):696-702.
[24] Efron N, Edwards K, Roper N, Pritchard N, Sampson GP, Shahidi AM, et al. *Eye Contact Lens* 2010 Sep;36(5):245-8.
[25] Papanas N, Ziegler D. Corneal *Current Diabetes Reports* 2013 Aug;13(4):488-99.
[26] Boulton AJ, Gries FA, Jervell JA. *Diabet Med* 1998 Jun;15(6):508-14.
[27] Feldman EL, Stevens MJ, Thomas PK, Brown MB, Canal N, Greene DA. *Diabetes Care* 1994 Nov;17(11):1281-9.
[28] Galloway MM. *Computer Graphics and Image Processing* 1975;4(2):172-9.
[29] Dasarathy BV, Holder EB. *Pattern Recognition Letters* 1991 Aug;12(8):497-502.
[30] Hall M, Franck E, Holmes G, Pfahringer B, Reutemann P, Witten IH. *SIGKDD Explorations* 2009;11(1).
[31] WLSVM: Integrating LibSVM into Weka Environment [computer program]. 2005.
[32] Vlachokosta AA, Asvestas PA, Gkrozou F, Lavasidis L, Matsopoulos GK, Paschopoulos M. *Med Biol Eng Comput* 2013 Aug;51(8):859-67.
[33] Hossain P, Sachdev A, Malik RA. *Lancet* 2005 Oct 15;366(9494):1340-3.
[34] Kallinikos P, Berhanu M, O'Donnell C, Boulton AJ, Efron N, Malik RA. *Invest Ophthalmol Vis Sci* 2004 Feb;45(2):418-22.

Tools For Interactive Learning And Self-Management Of Diabetes

Rita Capelo[1] **Carla Baptista**[2] **Júlia Figueiredo**[2]
Francisco Carrilho[2] **Pedro Furtado**[1]
[1]Departamento de Engenharia Informatica, [2]Universidade de Coimbra

E-mail: ritacapelo2@gmail.com, cfmbaptista@gmail.com, mariajuliateixeira@gmail.com, carrilhofm@gmail.com, pnf@dei.uc.pt

Abstract. Diabetes is a widespread disease and its control is dependent upon the patient. Although there is no permanent cure for diabetes, there are several available treatments which, when followed regularly, allow the patient to have a good quality of life. Patient education, especially about eating habits, is key to keep glucose levels stable both in the short and in the long term. This should include nutritional counselling, physical exercise, and the self-monitoring of glucose levels. The University of Coimbra and the Serviço de Endocrinologia, Diabetes e Metabolismo of Centro Hospitalar e Universitário de Coimbra started a collaboration to develop interactive tools for the learning and improvement of carbohydrate counting by patients. The approach presented in this paper is an interactive multimedia tool, available to patients through either the web or a smartphone. It helps them to learn how to maintain a healthy diet and how to monitor their insulin levels correctly by measuring the carbo-hidrate "equivalents" in meals. This application will create a more dynamic and interactive way of educating patients, improving solutions currently used in the Serviço de Endocrinologia, Diabetes e Metabolismo of the Centro Hospitalar e Universitário de Coimbra.

1. Introduction

Carbohydrate counting (CHC) is an established approach used by type 1 Diabetes patients to improve their glycemic control. However, CHC depends upon the patient's individual perception, and it may require a high level of subjectivity in the person's own self-assessment. Patients must learn how to recognize which foods contain CH and how to count the CH quantity in each of them. This requires a structured and continuous teaching, as well as an active participation of the patient. It is crucial to develop methods that minimize any errors the patient can make while calculating the CH amount of a specific food item or meal, since mistakes in determining these values can lead to an incorrect administration of prandial insulin.

We propose an interactive multimedia tool that offers both informative content, training, and serious games content that can help patients learn how to estimate the carbo-hydrate contents of meals. The application works by helping patients to achieve a better understanding of their disease, especially in

the counting of carbohydrate equivalents in food and meals. Thus patients can simplify the calculations they so often need to perform.
This paper is organized as follows: Section 2 contains related work; Section 3 describes the rationale behind the application that was created and its subsections describe each part of the application. Section 3.1 describes the Learning Area; Section 3.2 describes the Informational Area; Section 3.3 describes how the tools work; Section 3.4 describes the gaming area; finally, in Section 4 we discuss conclusions and future work.

2. Related Work

Some applications and resources aimed at helping diabetes patients are already available in the market. The ones which are most closely related to the subject discussed here can be found in several websites that are usually related to diabetes associations, charities, and companies that sell diabetes-related equipments and services.

The American Diabetes Association [1] provides important information for teaching patients how to control their glucose levels and how to administrate insulin. Their website provides video and text materials on the basics of diabetes, living with diabetes, and food and fitness. It includes videos and instructions on how to compose a plate (e.g. Six Easy Steps to Create Your Plate). It also provides food tips and elucidation on understanding carbohydrates.

The National Diabetes Education Program (NDEP) is a US partnership between the National Institutes of Health, the Centers for Disease Control and Prevention, and more than 200 public and private organizations [2]. This program provides free information about diabetes, using various learning resources and educational videos.

The Portal da Diabetes [3] includes information about diabetes, as well as a section of learning resources containing several brochures about different aspects of this disease. However, it does not have any interactive resources.

The mission of Learning About Diabetes, Inc. [4], is to provide easy-to-understand diabetes-care information to people with diabetes, the general public, and diabetes health care professionals. On their website we can find some learning materials about the diabetes in a quiz format. Its contents are purely static; they include both text and videos that explain what diabetes is.

There are many other organizations and companies working with diabetes which publish educational materials. For instance, MerkTM created a comprehensive program targeted at diabetes patients and healthcare professionals [5]. The program is called Journey for Control and brings together the key partners in the treatment of diabetes: Patients, Physicians, Diabetes Educators, and Health Planners. It is a comprehensive educational program designed to enlighten and empower patients by giving them the information and tools they need to make the lifestyle changes that lead to improved self-management. This website includes an explanatory guide about the disease and reference materials for its various components. OptumRx is a company specializing in the delivery, clinical management and affordability of prescription medications and consumer health products [6]. They publish materials related to educating about diabetes, such as downloadable documents and tips.

Joslin Diabetes Center [7] does present an interactive tool that allows you to drag food into a plate, like the one found in our application. It attempts to help the patient understand the nutritional values on each dish, but otherwise it does not provide the calculations that the patient must do in order to properly administer their insulin intake, and it does not include all the remaining interactive information that our application has.

The references reviewed above provide useful information for both patients and healthcare professionals. Our tool is an important addition to those mostly static or incomplete sources, as it adds interactivity and comprehensive meals and food search and visual interaction. Through it patients can experiment with plates; search for meals and types of foods in interactive ways; easily correlate what they found with carbohydrate counting, and patient must know many more features that allow them to achieve deeper insight. In particular, the gaming approach makes learning and experimenting intuitive and fun. By creating this tool, we aim to provide a set of interactive contents that the patient can interact with and through which he can assimilate knowledge, both in calculations and in the counting of "equivalents", a unit of measure frequently used in carbohydrate counting. The application contains both a learning component and interactive games that help to assimilate the relationship between food and carbohydrates and the insulin intake the patient must administer.

In addition, there are also smartphone applications designed to provide the user with information on food items, for example [8, 9, 10, 11, 12, 13]. These applications typically do not have most of the components we have created, including the gaming component, meals, equivalents, and carbon counting. Our application can also be deployed and used in smartphones.

3. Interactive Learning Experience with "Saber Contar, Saber Comer" (SCSC)

Our application is called "Saber Contar, Saber Comer" (SCSC), meaning "Know how to count, know what to eat". Our goal when creating this interactive application was to make it easier for patients to learn and to practice their CH counting skills. There are many different issues that need to be discussed in order to achieve this. In particular, it is necessary to provide the patient with general information, learning materials, and useful tools which they can use to search for information concerning specific meals (e.g. breakfast, lunch, mid-afternoon snack, dinner) and specific foods (e.g. types of bread, vegetables, meat, milk) and, most importantly, interactive serious games which the patient can use to test their knowledge and their guessing skills. In order to represent all these important features, our application is divided into four different areas, each containing four sections:

A. Learning

A.1. Correspondence between carbohydrates and equivalents;

A.2. Food label reading.

The Learning functionality aims to teach the patient how to match carbohydrate contents of meals to equivalents. An equivalent is the unit used to make carbohydrate counting easier. Furthermore, it also intends to teach the patient how to read different types of food labels. These two aspects are the key aspects every diabetes patient must to know in order to be able to count carbohydrates, control their insulin, and to better handle their illness.

B Information

B.1. Nutritional value of food items;

B.2. Meal suggestions.

This section allows the user to search for vast amounts of information, mostly pictorial in nature. It also provides the most important facts about carbohydrate counting for various meals and types of foods. Not only does the application allow for freeform searching, it also organizes the data into informative and logical types and categories. For instance, the patient will be able to find the contents of his breakfast (which may vary) and also to search for types of food, subtypes, and categories. This is further improved by allowing the user to enter quantities. The system will then provide the adequate CH counting, which can be learnt by the user.

C. Tools

C.1. Table of Individual data;

C.2. Insulin dosage calculator.

The calculation of insulin intake has individualized parameters, because of different sensitivity factors in each person. Our tools include customization for a specific patient, as well as a calculator of insulin intake that takes into account individual factors and the equivalents determined by the patient. This information is necessary in order for the patient to determine the adequate amount of insulin to take, or at least for the patient to learn how this is done.

D. Games

D.1. Equivalent counting

D.2. Insulin calculation

A serious game is a game that has an important practical training objective beyond just having fun. Serious games are especially appropriate for educating diabetic patients. These games are all based on imaging data. First, the user drags food and drinks into plates and glasses, respectively. Then they can raise or lower the quantities by dragging more or less items into the plate or glass or by clicking up or down. This makes the games very intuitive and easy to use by anyone, from children to old people. Its ease of use and intuitiveness make the tool quite appropriate for our purpose. We have included features for counting equivalents – the patient may compose a plate with any food he likes, all virtually, by dragging and dropping – and for insulin intake calculation, as an added feature.

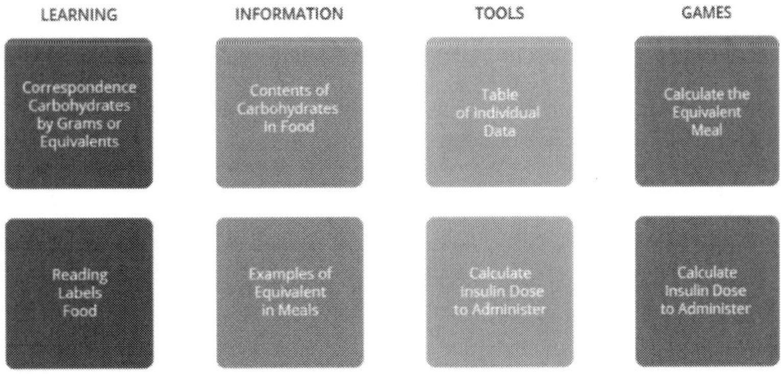

Figure 1. "Saber Contar Saber Comer" Main Menu.

Next we describe each part of the application in more detail, and provide illustrations for each. Section 3.1 describes the Learning Area, section 3.2 describes the Informational Area, section 3.3 describes how the tools work and section 3.4 describes the Gaming area.

Journal of Physics: Conference Series **616** (2015) 012003 doi:10.1088/1742-6596/616/1/012003

3.1. Learning Area

The "Learning" area, shown in Figure 2, is meant to educate the patient. It includes an explanation of how carbohydrates are converted into equivalents. It also teaches patients how to read food labels, and contains three examples of labels in food items chosen by the hospital for their convenience. These are yogurt, cookies, and breakfast cereal.

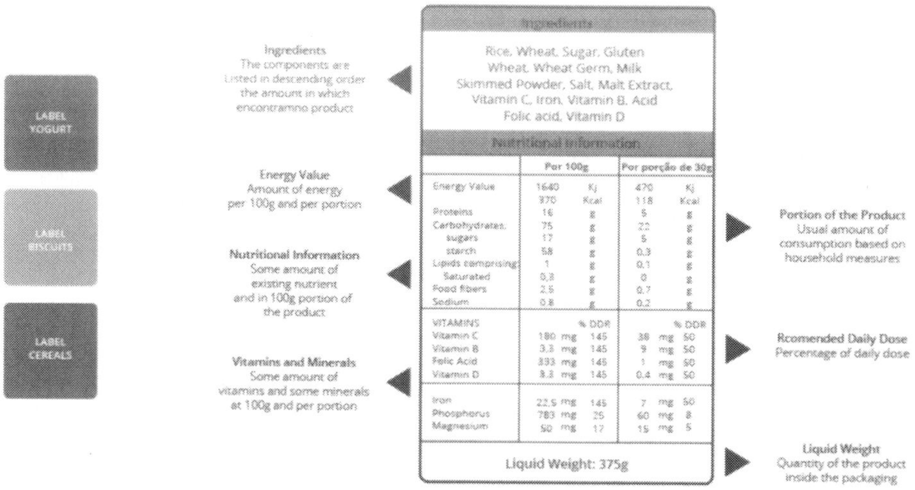

Figure 2. Food label reading

3.2. Information Area

The "Information" area is a tool that allows the user to find out how many equivalents exist in a food item. The user can choose between the following food categories: Starches, Fruits, Vegetables, Dairy, Soups, and Candy and Deserts. After selecting a category, the user sees a horizontal scroll with pictures of the more commonly eaten foods. Furthermore, on the bottom of the area they can find a search function, which can be used to find a food item that does not appear on the above list. After choosing a certain food and entering the desired quantity, the user sees the number of equivalents, calories, and its protein and carbohydrate amounts. The second section displays examples of menus for each meal of the day, in order to help the patient to follow a healthy diet.

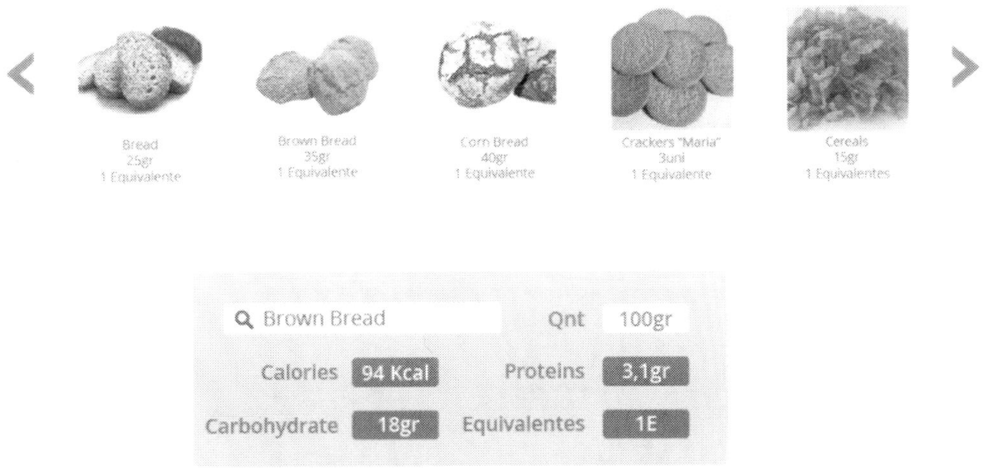

Figure 3. Nutritional value of food items

Figure 4. Meal suggestions

3.3.Tools Area

The "Tool" area includes two pages. The first one includes a table where the user can enter their personal data (as prescribed by their doctor). This is very important in order to learn the specific insulin dosage to be administered after each meal. The second page contains a calculator tool that helps the patient to perform calculations by introducing their personal data.

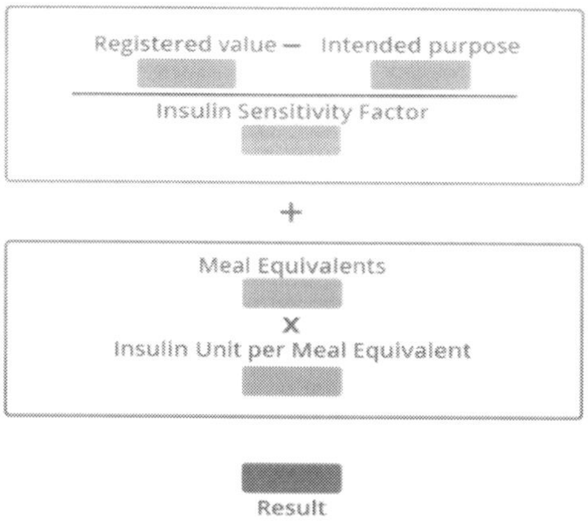

Figure 5. Insulin dose calculator

3.4. Games Area

Finally, the Gaming area includes two games which, despite their graphic similarities, have completely different goals. In the first game the user has to calculate the equivalents in a pictured food dish, based on what they have learnt in medical appointments. In the second game, the user has to calculate the insulin dosage they must administer, based on the dish pictured and their own insulin levels. These games allow users to drag food items into their plate and thus understand the amount of carbohydrates or equivalents they are seeing. Therefore, the games are an entertainment component of the application that also help patients to assimilate important information.

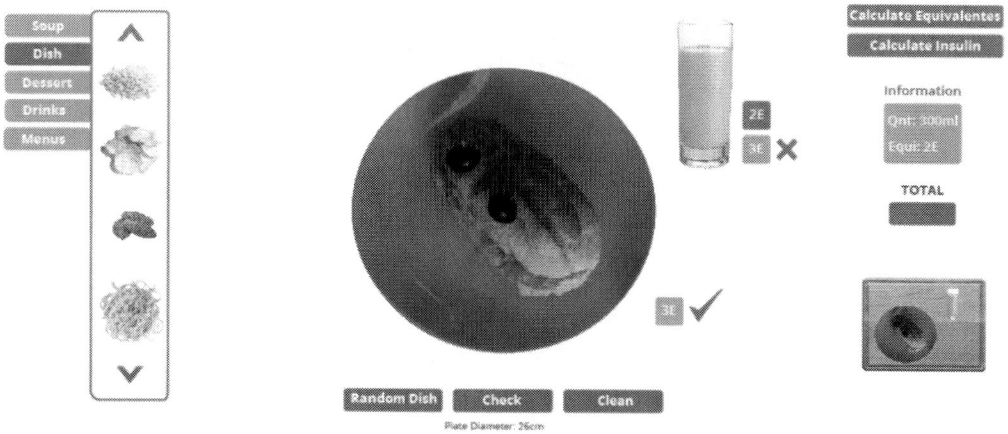

Figure 6. Interactive insulin calculator game area

4. Conclusions

In this paper we have presented a new approach aimed at teaching and training diabetes patients in carbo-hydrate counting and insulin intake planning. This interactive tool was developed in collaboration with the Servico de Diabetes e Metabolismo of CHUC Hospital. We discussed the purpose of creating this application and we explained how important it is for diabetes patients to learn how to measure their insulin intake appropriately, and why they must calculate the amounts of carbo-hydrates in meals correctly.

Currently, there are no available applications in the market that help diabetes patients to learn what to eat and how to count carbohydrates in this way. Although there are many interesting diabetes educational applications elsewhere, the interactivity that is proposed in our approach is unique and can highly benefit patients by improving their quality of life in a significant way. Unlike other approaches, this application empowers the diabetes patient by teaching him how to become independent. It both allows patients to assimilate information they have learned in their medical appointments and also teaches them how to apply this information, using the application's interactive game area.

5. References

[1] American Diabetes Association Website http://www.diabetes.org/diabetes-basics/ [Accessed in 11/2014]

[2] National Diabetes education program - http://ndep.nih.gov/diabetes-facts/index.aspx [Accessed in 11/2014]

[3] Portal da Diabetes - http://www.apdp.pt/ [Accessed in 11/2014]

[4] Learning About Diabetes - http://www.learningaboutdiabetes.org [Accessed in 11/2014]

[5] Merck & Co. http://educator.journeyforcontrol.com/diabetes_educator/professional_resources/ [Accessed in 10/2014]

[6] OptumRx - https://www.optumrx.com/diabetes_testing_supplies.html [Accessed in 10/2014]

[7] Joslin Diabetes Center - http://www.joslin.org/healthy-plate/demo.html [Accessed in 11/2014]

[8] Aplicação Social Diabetes - https://play.google.com/store/apps/details?id=com.socialdiabetes.android [Accessed in 09/2014]

[9] Aplicação Social Diabetes - https://play.google.com/store/apps/details?id=com.socialdiabetes.android [Accessed in 09/2014]

[10] Aplicação HelpDiabetes - https://play.google.com/store/apps/details?id=com.hippoandfriends.helpdiabetes [Accessed in 09/2014]

[11] Aplicação Gluci-Check - https://play.google.com/store/apps/details?id=com.roche.glucichek [Accessed in 09/2014]

Journal of Physics: Conference Series **616** (2015) 012003 doi:10.1088/1742-6596/616/1/012003

[12] Aplicação Carbs&Cals - https://play.google.com/store/apps/details?id=com.chello.carbsandcals&hl=pt_PT [Accessed in 09/2014]

[13] Aplicação GliControl - https://play.google.com/store/apps/details?id=air.com.softbycloud.glicontrol&hl=pt_PT [Accessed in 09/2014]

Population Accessibility to Radiotherapy Services in New South Wales Region of Australia: a methodological contribution

Nagesh Shukla[1,3], Rohan Wickramasuriya[1], Andrew Miller[1,2], Pascal Perez[1]
[1] SMART Infrastructure Facility, Engineering and Information Sciences, University of Wollongong, Wollongong, NSW 2522, Australia
[2] Illawarra Cancer Care Centre, Wollongong Hospital, Wollongong NSW 2500 Australia

E-mail: nshukla@uow.edu.au

Abstract. This paper proposes an integrated modelling process to assess the population accessibility to radiotherapy treatment services in future based on future cancer incidence and road network-based accessibility. Previous research efforts assessed travel distance/time barriers affecting access to cancer treatment services, as well as epidemiological studies that showed that cancer incidence rates vary with population demography. It is established that travel distances to treatment centres and demographic profiles of the accessible regions greatly influence the demand for cancer radiotherapy (RT) services. However, an integrated service planning approach that combines spatially-explicit cancer incidence projections, and the RT services accessibility based on patient road network have never been attempted. This research work presents this novel methodology for the accessibility assessment of RT services and demonstrates its viability by modelling New South Wales (NSW) cancer incidence rates for different age-sex groups based on observed cancer incidence trends; estimating the road network-based access to current NSW treatment centres; and, projecting the demand for RT services in New South Wales, Australia from year 2011 to 2026.

1. Introduction

Cancer control is a health priority in almost every national jurisdiction. Cancer is estimated to be the leading cause of burden of disease in Australia in 2010, accounting for 19% of the total burden [1]. This becomes increasingly important as life expectancy continues to grow; the proportion of elderly people in the population will steadily increase over the next decades [2]. Therefore, it is expected that the number of cancer cases will continue to grow, as the 'baby boomers' ageing population is entering the high incidence period.

Beyond demographic influences, other factors like socio-economic status and ethnicity have also an effect on cancer incidence [2]. Furthermore, geographical variations also occur among the rate of treatment and survival from cancer [3-4]. As the number and diversity of cancer cases increase, the pressure on specialised treatment services will increase as well, calling for better planning and allocation of healthcare resources, particularly at the regional level.

[3] Corresponding Author

Radiotherapy (RT) is an essential cancer therapy whether aimed at cure or palliation. The Collaboration for Cancer Outcomes Research and Evaluation (CCORE) literature-based findings were used by the Radiation Oncology Reform Implementation Committee (RORIC) to estimate that 52.3% of all diagnosed cancer cases in Australia would benefit from radiotherapy at some point after diagnosis. The Australian state health department for New South Wales (NSW) uses this estimate for planning their RT services [5-7].

Similar to other jurisdictions, NSW studies of access to RT services have established that utilization rates are well below this optimal number [8-11] clearly demonstrating the existence of barriers to accessibility to existing RT services. Currently, only 38% of cancer sufferers receive radiotherapy at some point after the initial detection [2,12-13] indicating that ~14% of cancer sufferers, who would benefit from RT services, miss the opportunity for improved cancer control and appropriate evidence-based management [2,12,14].

Various studies have examined the effect of geographical accessibility, based on travel times/distances as proxy to travel effort, on uptake of RT based cancer treatment. For example, Madelaine et al. [15] reported lower treatment rates for rural lung cancer patients in France. Punglia et al. [16] found that increasing distance to the nearest radiotherapy centre was associated with a decreasing likelihood of receiving post-mastectomy radiation therapy. Greenberg et al., [17] asserted that lung cancer patients living at greater straight line distance from a specialist cancer centre, in rural USA, were significantly more likely to undergo surgery but were less likely to receive radiotherapy or chemotherapy than closer patients. Athas et al. [11] also found that breast cancer patients living further than 75 miles from a radiotherapy services centre were significantly less likely to receive radiotherapy than those living closer. It is possible that the deterring effect of transportation may be even more pronounced in patients who are faced with weeks of daily outpatient treatment, as is common for radiation therapy.

The planning of efficient and accessible RT services for cancer care at regional level requires appropriate estimates of current and future cancer demand based on the spatial distribution and evolution of various socio-demographic groups, spatial accessibility based on transport network and probabilities of re-treatment. In this study, we will develop a modelling tool which can be systematically used for planning of radiotherapy services.

2. Methodology for planning of radiotherapy services

The proposed methodology for modelling and predicting the future cancer incidences and their accessibility to existing RT centres in the state involves datasets such as Australian Institute of Health and Welfare (AIHW) cancer incidence data, Australian Bureau of Statistics (ABS) population projection dataset, residential areas dataset, road network dataset, and data about the existing radiotherapy (RT) centres.

2.1. Datasets used in the proposed methodology

2.1.1. AIHW cancer incidence dataset

The AIHW cancer incidence dataset [18] provides the information about age group and sex specific cancer rates for all and specific cancer types in Australia. The major source of this data comes from the Australian Cancer Database (ACD) which contains records of all the primary, malignant cancers (except basal cell and squamous cell carcinomas of the skin) diagnosed in Australia since 1982. The age-specific rates (ASR) for cancer in males and females, in past years, are used to model and predict the cancer incidences in the state.

2.1.2. ABS population projection dataset

The population projection data prepared by the ABS to permit the Australian Government Department of Social Services to plan and evaluate aged care. These projections are based on the past trends (over a decade) of fertility, mortality, and migration trends, using the cohort-component method, where the base population is projected into the future year annually by estimating the effect of births, deaths and migration within each age-sex group according to the specified fertility, mortality and migration assumptions. The population projections cover the period from 2012 to 2026, using the 2011 ABS census year as a base. More information about this dataset is presented elsewhere [19].

2.1.3. Residential areas dataset
A digital spatial dataset of the residential areas in NSW (Figure 2) is extracted from the catchment-scale land use maps collated by the Australian Collaborative Land Use and Management Program (ACLUMP), which is established to promote the development of nationally consistent land use and land management practices information for Australia.

2.1.4. Road network dataset
The road network of NSW was downloaded from the most up-to-date OpenStreetMap (OSM) database, using Quantum GIS software [21]. OpenStreetMap is a crowd-sourced initiative to collect and map roads, trails, and points of interest, with an ultimate aim of building a geographic database that contains every single feature on the planet [22]. OSM data was selected for this study as it is a readily available and open data source with sufficient positional accuracy [23].

2.1.5. Existing radiotherapy (RT) centres
The data about the existing RT treatment facilities is accessed from Department of Health, Australia. DoH [24] provides publically available information about radiation therapy services including accommodation and travel schemes, facility locations and treatment options. The facility locations for RT centres were used for estimating the RT treatment accessible regions for future cancer incidences.

2.2. Methods

2.2.1. Age specific rate (ASR) for cancer incidence modelling
Ordinary least squares linear regression is used to model the past trend of the cancer incidence. The linear regression models have been developed for each age-sex group. Age specific rates (ASR) from AIHW dataset, for cancer incidences for 10 years (2000-2009), have been used to estimate the coefficients of the linear regression model. Time in years is used as a variable to determine the ASR in each age-sex group.

The confidence level is set to be 95% for the two sided hypothesis test. If the ASR model for an age-sex model is found significant, then the rate is estimated based on the regression model. Otherwise, mean of ASR (for years 2000 – 2009) for a particular age-sex group is used for prediction.

2.2.2. Predicting the cancer cases
Once the regression models for the cancer incidence is obtained for each age-sex group, they are then used together with the ABS population projections dataset to estimate the future cancer cases in the study area. The ABS population dataset is aggregated into a contingency table having attributes such as *age group, sex, geographical area*, and *number of people at risk*. This data table is then used together with the predicted ASR for each year (derived from section 2.2.1) to produce *cancer cases* in different age-sex group in different local government areas (LGAs). The result of this analysis provides cancer cases distributed in various LGAs in future. Once the number of cancer cases for future years are identified, it can be used to estimate the future radiotherapy demand/workload.

2.2.3. Travel distance modelling and the estimation of accessible residential area
Following the literature that supports a strong relationship between travel distance and uptake rates (as discussed in Section 1) we calculated constant driving distance polygons using the popular open source software 'pgRouting' [25], taking treatment centres as points of origin.

Routing functions provided by pgRouting were implemented as an extension to the open source PostgreSQL/PostGIS geospatial database [26], and the OSM road network of NSW was imported into PostgreSQL/PostGIS as a pgRouting-enabled (i.e. routable) network dataset. A custom function that leverages the default pgRouting function 'pgr_drivingDistance' was used to loop through all treatment centres, thereby calculating constant driving distance polygons for all centres and for several distances. The function 'pgr_drivingDistance' utilises Dijkstra shortest path solution to calculate constant polygons [25]. When a polygon calculated for a large driving distance encompasses a polygon for a smaller driving distance, the area of the smaller polygon is then erased from the larger polygon to construct a ring for a range of driving distances (e.g. from 50km to 100km).

Considering the fact that dwellings are not uniformly distributed in space, the digital dataset of the residential areas of NSW is used to estimate the fraction of the population that live in a particular LGA within a given distance ring.

2.2.4. Radiotherapy based treatment demand estimation

The predicted cancer cases and ratios of LGA within each distance bands were used together with the radiotherapy rates to estimate the future demand for RT services. The RT rates were chosen based on the distance bands provided in Gabriel et al. [27]. Furthermore, Delaney et al. [28] have summarised the overall optimal radiotherapy utilization rate for all registered cancers in Australia. They summarised the recommended optimal radiotherapy utilisation rates for individual cancers such as skin, breast, lung, gastrointestinal, genitourinary and others. Evidence for radiotherapy based treatments such as systematic review of relevant randomised studies, randomised/controlled trials, and case series, were used in modelling decision trees for determining optimal radiotherapy rates. For more information about these rates and methodology see [28]. The optimal RT rates varied from 0% in case of liver cancer patients and 92% for patients with central nervous system tumors. The overall estimate of optimal radiotherapy utilisation rate was estimated to be 52.3% of all the notifiable cancer cases in Australia. Furthermore, an additional 23% of the cancer patients who have already received RT will require retreatment according to the SSDB report [29]. This means that an apparent 64.33% (52.3% + 0.23×52.3%) of the total cancer patients would need radiotherapy treatments in future. We have used this estimated rate of 64.33% for determining the optimal RT demand for the future year.

All the methods in this section are coded in R, which is an open source software environment for statistical computing and graphics using open source R libraries such as *maptools, ggplot2, plyr, rgeos, rgdal, sp, RColorBrewer* for this methodology development and data analysis.

3. Results and Discussion

This study estimates a spatially explicit demand and accessibility for radiotherapy services in the NSW state of Australia using the methodology described above. The choice of the NSW was justified by the accurate data currently available for NSW and a well-defined geographical area that is known to mirror average demographic distributions in Australia.

Ordinary least squares linear regression for each age-sex group were conducted on the past cancer incidence data from AIHW. In females, the incidence rates for age groups *'55-59 yrs'*, *'65-69 yrs'*, and *'80-84 yrs'* were found significant at 95% confidence interval. In case of males, the incidence rates for age groups *'45-49 yrs'*, *'50-54 yrs'*, *'55-59 yrs'*, *'60-64 yrs'*, *'65-69 yrs'*, and *'70-74 yrs'* were found significant at 95% confidence interval. Figure 4 illustrates the results from linear regression analysis (all age groups combined) of cancer incidences projected in future and its comparison with the past incidence rates. In Figure 1, the overall rates for cancer incidence in females are not increasing compared to the incidences in males. The dotted lines in Figure 1 represent the results of the model and the solid lines represent the actual rates observed. The results of the ASR prediction model for future years were then applied to the ABS population projections.

The NSW population data provided by the ABS was used in this study. It is estimated that the population will grow significantly in future. However, the growth in population in different age groups and sex varies across NSW. Figure 2 illustrates the growth of population in different age-sex group in year 2011 and 2026. In year 2026, number of individuals in age groups with a higher cancer incidence (i.e., above 50 years) will be larger than 2011 for both females and males. There will be more aged people living in year 2026 in NSW, and because of cancer incidence in the aged population sub-groups, more cancer cases.

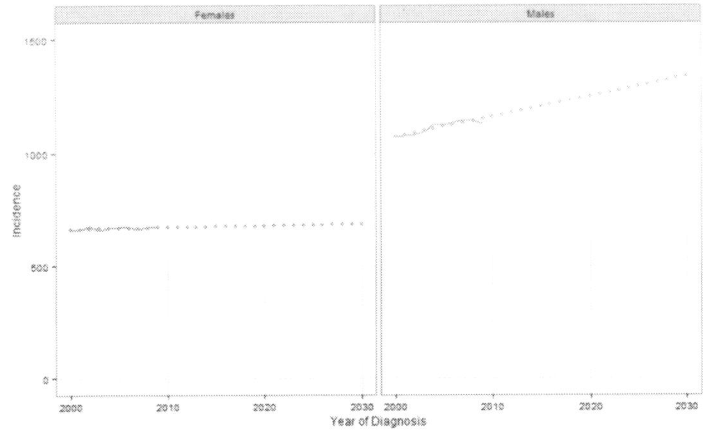

Figure 1. Predicted (points) and observed (solid line) incidence rates (per 100,000) for all cancers in males and females in Australia

Figures 3a and 3b illustrate the cancer incidences (cancer cases per 100,000 individuals) for year 2011 (Figure 3a) and future year 2026 (Figure 3b). The cancer incidences in all the local government area (LGA) of NSW (total of 154) are based on the age-sex specific rates obtained based on ordinary least square linear regression modelling ($p<0.05$, see Figure 1) and future population estimates data (see Figure 2). In year 2011, the average cancer incidence at LGA level is approximately 617 per 100,000 individuals. However, it increases to 812 (approx.) per 100,000 individuals for future year 2026. Based on the spatial distribution of cancer incidences in year 2011 and 2026, which suggests more cancer cases in the coastal, mid-eastern and southern LGAs of NSW, planning of radiotherapy services for these areas is important.

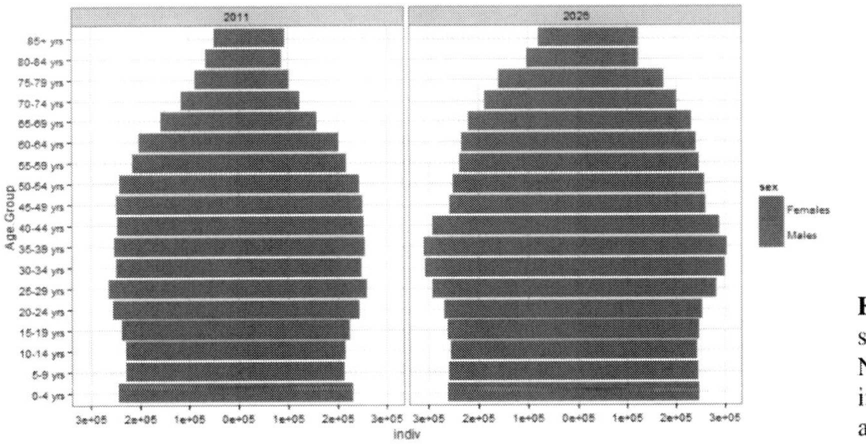

Figure 2. Age structure of NSW population in years 2011 and 2026

(a)

(b)

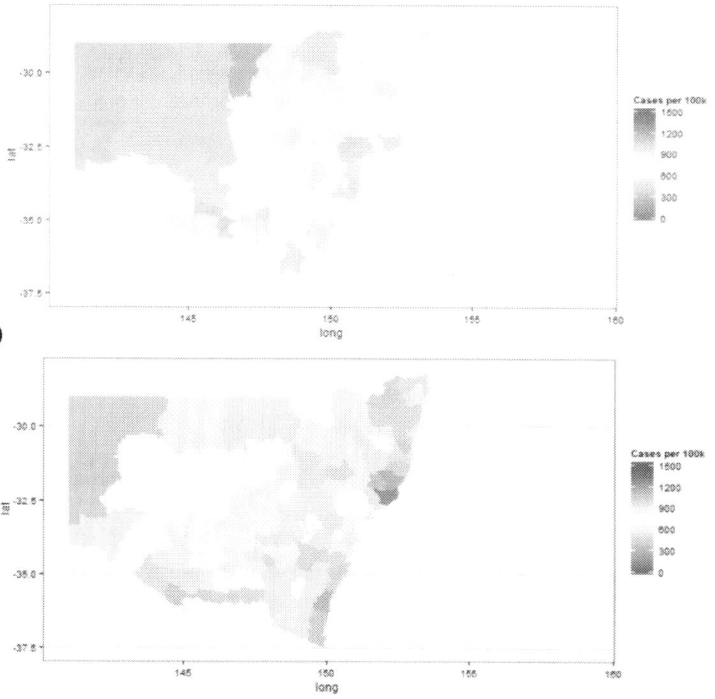

Figure 3. Overall cancer incidences in year 2011 (a) and 2026 (b) in NSW state of Australia

Figure 4 depicts driving distance polygons or the accessible regions considering all radiotherapy centres as trip origins. We have included existing radiotherapy centres in the NSW and also the nearby centres from the two bordering states, Victoria and Queensland. This is done to account for patient leakage in to those centres from NSW. As described in Section 2.2.3, for each LGA the fraction of population living in each distance band is estimated. Table 1 shows these calculated fractions for two LGAs as an example.

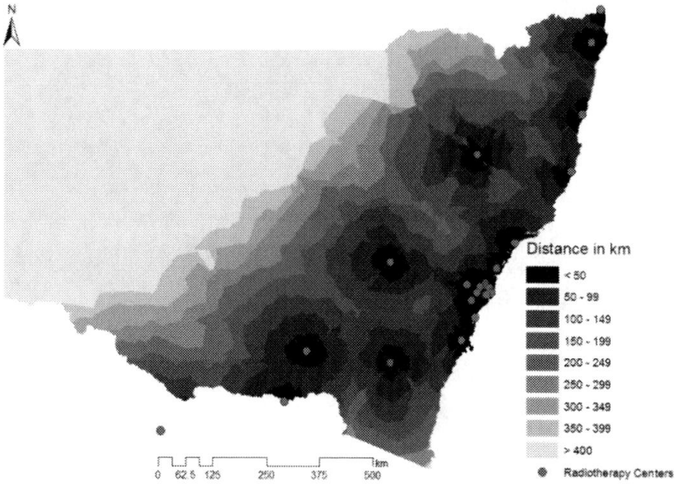

Figure 4. Constant driving distance polygons from radiotherapy centres

Table 1. Estimated fractions of residential population living in constant driving distance bands from radiotherapy centres

LGA	Distance band	Fraction of population
Hawkesbury	< 50	0.88
Hawkesbury	50 - 99	0.11
Hawkesbury	100 - 149	0.01
Wingecarribee	< 50	0.15
Wingecarribee	50 - 99	0.85

The ratios for each LGA in the NSW (illustrated in Table 1) were used together with the cancer cases for respective LGA (illustrated in Figure 3) to estimate the total number of cancer patients residing in each distance band. The cancer cases residing in each distance band can be used for planning and locating new radiotherapy centres.

Table 2 illustrates the demand for the RT services based on the above mentioned modelling methods. RT demand at the current service level is based on the RT rates by road distance estimated in Gabriel et al. (2012). However, the optimal RT demand is estimated by using cancer incidence number and optimal RT utilisation rates (64.33%).

Table 2: RT demand from year 2011 to 2026

Year	RT Demand at current service level	Optimal RT demand based on evidence
2011	10898	21665
2016	12602	25056
2021	14472	28755
2026	16452	32651

4. Conclusion

The treatment of cancer, which touches a large proportion of the community, represents a significant health and economic burden in Australia. Nearly one in two men and one in three women have a lifetime risk of being diagnosed with cancer in NSW. Radiotherapy (RT) based cancer treatments are often integral to achieving permanent or long-term remission. There will be large demand for RT services for cancer patients in NSW in next ten years. However, decision on new LINAC units usually take a long time to be made, resulting in local congestion of services or unreasonable travel time for patients, with some individuals opting out of RT treatment. Thus, there is pressing need to estimate patient's accessibility to respective treatment centres in future. This can also help to rationally justify the placement of new radiotherapy centres to meet the future RT demand. This study makes a unique contribution to the policy debate on accessibility to RT treatment centres through the establishment of an integrated modelling approach. This study marks an attempt at informing the government about maximising accessibility to RT centres to meet future cancer patient demand.

The demonstrated generic methodology includes cancer incidence predictive modelling to estimate the cancer cases residing within different travel distance bands from the RT centres, and is easily applied to the NSW circumstance. The methods described in this research study can be applied in other jurisdictions with availability of the necessary data.

5. References

[1] Cancer Council Australia. Facts and Figures. Sydney: Cancer Council Australia (2012)

[2] ROTC (Radiation Oncology Tripartite Committee), (2012a), Planning For The Best: Tripartite National Strategic Plan for Radiation Oncology 2012 – 2022, The Royal Australian and New Zealand College of Radiologists

[3] Coory, M.D., Ho, T. and Jordan, S. J., Australia is continuing to make progress against cancer, but the regional and remote disadvantage remains, *Med J Aust* 2013; 199 (9): 605-608.

[4] Yu, X.Q., Luo, Q., Smith, D. P., O'Connell, D.L. and Baade, P.D., Geographic variation in prostate cancer survival in New South Wales, *Med J Aust* 2014; 200 (10): 586-590.

[5] Delaney GP., Jacob S., Featherstone C., Barton MB. Radiotherapy in Cancer Care: Estimating Optimal Utilisation from a Review of Evidence-Based Clinical Guidelines. Collaboration for Cancer Outcomes Research and Evaluation (CCORE), Liverpool Hospital, Sydney, Australia, 2003.

[6] ROTC (Radiation Oncology Tripartite Committee), (2012b),Projecting the radiation oncology workforce, Input to the Tripartite National Strategic Plan For Radiation Oncology in Australia, May 2012, The Royal Australian and New Zealand College of Radiologists

[7] Morgan G. Why has Radiotherapy Utilisation not improved since 1999? *Journal of Medical Imaging and Radiation Oncology*. 2011 August; Volume 55 (Number 4) p347-350

[8] Mackillop WJ, Groome PA, Zhang-Solomons J et al. Does a centralized radiotherapy system provide adequate access to care? *J ClinOncol* 1997; 15 : 1261–71.

[9] Barton M. Radiotherapy utilization in New South Wales from 1996 to 1998.*AustralasRadiol* 2000; 44: 308–14.

[10] Einhorn J, Frödin J-E, Möller T et al . Radiotherapy practices in Sweden compared to the scientific evidence. *Acta Oncol* 1996; 35 (Suppl. 6): 89–97

[11] Athas WF, Adams-Cameron M, Hunt WC, Amir-Fazli A, Key CR. Travel distance to radiation therapy and receipt of radiotherapy following breast-conserving surgery. *J Natl Cancer Inst,* 2000, 92, 269–71.

[12] Barton M, Delaney G. Decade of investment in radiotherapy in New South Wales: why does the gap between optimal and actual persists? *Journal of Medical Imaging and Radiation Oncology*. Volume 55, Issue 4, August 2011, pp: 433–441

[13] Denham JW. How do we bring an acceptable level of radiotherapy services to a dispersed population? *Australas Radiol* 1995; 39: 171–3.

[14] NSW Health Dept. 2001, Statewide Services Development Branch. NSW Radiotherapy Management Information System. 2001. NSW Health Department

[15] Madelaine J, Guizard AV, Lefevre H, Lecarpentier MM, Launoy G. Diagnosis, treatment and prognosis of lung cancer in the Manche (France) (1997 - 1999) according to patient socioeconomic characteristics, *Review Epidemiol.* Sante Publique 2002;50:383–92.

[16] Punglia RS, Weeks JC, Neville BA, Earle CC. (2006), Effect of distance to radiation treatment facility on use of radiation therapy after mastectomy in elderly women. *Int J Rad Oncol Bio Phys* 2006; 66:56–63.

[17] Greenberg ER, Chute CG, Stukel T, et al. Social and economic factors in the choice of lung cancer treatment.A population-based study in two rural areas. *New Engl J Med* 1988;318:612–7.

[18] AIHW (2014), Cancer Data, (Accessed Jun 2014) http://www.aihw.gov.au/cancer-data/

[19] DoH (2013), Statistical Local Area Population Projections, 2011 (base) to 2026, Preliminary, (Accessed, Jun 2014): http://www.health.gov.au/internet/main/publishing.nsf/Content/ageing-stats-lapp.htm

[20] ALUM (2010). The Australian Land Use and Management (ALUM) Classification Version 7, May 2010, (Accessed, August 2014): http://www.daff.gov.au/ABARES/aclump/Documents/ALUM_Classification_V7_May_2010_detailed.pdf

[21] Gray, J (2008), Getting Started with Quantum GIS. LINUX Journal, (Accessed, August 2014): http://www.linuxjournal.com/content/getting-started-quantum-gis

[22] Bennett J (2010), OpenStreetMap, Packt Publishing, Birmingham, GBR. p. 252.

[23] Haklay, M (2010), How good is volunteered geographical information? A comparative study of OpenStreetMap and ordnance survey datasets, *Environment and Planning B: Planning and Design*, Vol. 37, No. 4, pp. 682-703.
[24] DoH (2014), Radiation Oncology and Radiotherapy Services, (Accessed, Jun 2014): http://www.health.gov.au/internet/main/publishing.nsf/Content/health-roi-radiother-index.htm
[25] pgRouting Contributors (2013), pgRouting Manual, Release 2.0.0, p. 88, (Accessed , August 2014): http://docs.pgrouting.org/2.0/en/pgRoutingDocumentation.pdf
[26] Douglas, K, Douglas, S (2005), PostgreSQL, SAMS Technical Publishing, Indianapolis, IN, p. 1032.
[27] Gabriel, G., Barton, M., Delaney, G. Radiotherapy utilisation rates in NSW and ACT [2004-06], a data linkage and GIS experience. *Innovation in Cancer Treatment and Care NSW Conference*, 2012.
[28] Delaney G "Radiotherapy in Cancer Care: Estimating the Optimal Utilization from a review of Evidence-Based Clinical Guidelines" *PhD Thesis*, 2007.
[29] SSDB (Statewide Services Development Branch, 2004). Radiotherapy management information system report 2003. Sydney: NSW Health Department, 2004

Implementation of Advanced Warehouses in a Hospital Environment – Case study

J Costa [1], M Sameiro Carvalho [2], A Nobre [3]
1 Algoritmi Research Center, University of Minho, Portugal
2 Algoritmi Research Center, University of Minho, Portugal
3 Logistics Manager, Hospital de Braga, Portugal

E-mail: jcosta87@gmail.com

Abstract. In Portugal, there is an increase of costs in the healthcare sector due to several factors such as the aging of the population, the increased demand for health care services and the increasing investment in new technologies. Thus, there is a need to reduce costs, by presenting the effective and efficient management of logistics supply systems with enormous potential to achieve savings in health care organizations without compromising the quality of the provided service, which is a critical factor, in this type of sector. In this research project the implementation of Advanced Warehouses has been studied, in the Hospital de Braga patient care units, based in a mix of replenishment systems approaches: the par level system, the two bin system and the consignment model. The logistics supply process is supported by information technology (IT), allowing a proactive replacement of products, based on the hospital services consumption records. The case study was developed in two patient care units, in order to study the impact of the operation of the three replenishment systems. Results showed that an important inventory holding costs reduction can be achieved in the patient care unit warehouses while increasing the service level and increasing control of incoming and stored materials with less human resources. The main conclusion of this work illustrates the possibility of operating multiple replenishment models, according to the types of materials that healthcare organizations deal with, so that they are able to provide quality health care services at a reduced cost and economically sustainable. The adoption of adequate IT has been shown critical for the success of the project.

1. Introduction

Recent studies show that the healthcare sector is going through great changes with healthcare organizations suffering huge pressures to reduce costs and to increase healthcare quality (de Vries, 2011; Lee, et al., 2011). The healthcare costs in Portugal are increasing, essentially as a result of three factors: the aging population, new technology investments (and the associated labor costs) and the increased demand for healthcare (Pinto, 2008). Besides that, the efficiency of the National Health System is low and, the financial debt is high, even though it has been decreasing lately (Carvalho, 2010).

To provide healthcare services was for years the main concern of healthcare organizations, regardless of the associated costs. With the increase of costs associated with material managing and due to the severe competition between healthcare institutions, the importance of the Supply Chain Management (SCM) increased focusing on maintaining a good cost-efficiency ratio without compromising the service quality (Callender & Grasman, 2010; Carvalho, 2010). The main focus of the SCM is to reduce the associated costs of material and information flow management without reducing the service quality, acting as a critical factor on its organizational performance and on gaining advantage towards competitors (Lee, et al., 2011). Although the SCM practices have been proven effective in other industries, with considerable improvements, the results on the healthcare sector are far from the same level of improvement (McKone-Sweet, 2005; Rosseti, 2008; Kelle &

Schneider, 2012). For many authors, healthcare organizations cannot reduce costs due to inefficiencies in the SCM (Schneller & Smeltzer, 2006; de Vries, 2011). In the past few years the large number of studies performed on economics, logistics, operational research and administration, has contributed to develop and improve the healthcare sector. These studies show an increased concern with the inventory management since, drugs and medical consumables, typically accounts for the second largest expense in the hierarchy of hospital's costs. Therefore, special attention must be given to materials management, with a huge potential intervention in terms of logistics management (Schneller & Smeltzer, 2006; Nabais, 2009; Carvalho & Ramos, 2010; Landry & Beaulieu, 2010b).

This paper shows the results of a study conducted at Hospital de Braga (HB), in order to improve the process of material supply in patient care units, increasing the visibility of material flow, reducing inventory levels, and simultaneously increasing the material's availability. The project involved the reorganization of processes in the Central Warehouse (CW) and the implementation of the Advanced Warehouse (AW) concept for two clinical departments: Angiography and Central Operating Room (COR). The new system which integrates three different supply models (par level system, two-bin system and consignment system) for different types of materials, is supported by an information technology (IT) system, which allows users to know, in real time, the stock of material along the whole process, and to ensure its efficient management.

The paper is organized in six chapters: the next one is a theoretical framework on approaches to healthcare organization's logistic system; Chapter 3 describes the case study; Chapter 4 describes the implementation of the AW concept in the HB; in chapter 5, the results will be analyzed and discussed regarding the impact of this implementation; The main conclusions are described in the last chapter.

2. Literature review

According to the Council of Supply Chain Management Professionals SCM "includes the planning and management of all activities involved in sourcing, procurement, conversion, and logistics management. It should be noted that the SCM involves the coordination and collaboration between partners, whether they are suppliers, logistics providers or customers. In essence, the SCM integrates supply and demand management within and across companies" (Carvalho, 2010).

The SCM in hospitals includes the external supply chain (e.g. suppliers, distributors, etc.) and the internal supply chain (e.g. hospital facilities, warehouses, patients, etc.) (Landry & Beaulieu, 2002; Rivard-Royer, 2002; Schneller & Smeltzer, 2006). A hospital receives goods and services from suppliers and then stores and distributes them to various patient care units, according to the operating process. Internally, the hospital has a complex distribution network, which usually consists of warehouses that provide stocks of medical supplies to patient care units (Landry & Beaulieu, 2002).

The specific context of the healthcare sector differs from the industrial sector, as its main mission is to provide healthcare services to patients (Landry & Beaulieu, 2010a). However, to provide healthcare services to patients, healthcare institutions need material resources. These institutions need to make significant investments in material resources to ensure proper healthcare services. Effective management of materials can lead to huge savings in these organizations, while the definition of robust stock management policies is essential to deal with the complexity associated with materials consumed (Carvalho & Ramos, 2010). Recent studies show that improving the stock management results in increased savings and revenues (McKone-Sweet, 2005; Rossetti, 2008).

Inefficient management of stocks can cause excess of or material shortage, leading to consequences for the healthcare organization. On one hand, excess inventory conducts to high holding costs, which leads to the risk of products becoming obsolete. On the other hand, the lack of inventory may result in sotck-outs, which in the healthcare industry can mean dissatisfaction among healthcare professionals, delays in surgery, and in extreme situations, in the death of patients (de Vries, 2011). In these circumstances it is normal healthcare professionals wanting large amounts of stocks (Landry & Beaulieu, 2010a). Many healthcare organizations remain hesitant when it comes to reducing inventory levels, since the costs associated with the lack of material are much higher than the cost of having additional inventory (Aptel & Pourjalali, 2001).

The distribution of materials to patient care units of healthcare organizations is usually performed in two ways: the distributor delivers the material in the hospital's CW, and the CW is responsible for storing and delivering the material to the patient care units; or the distributor (or CW after receiving) delivers the material directly to the patient care units, and there is no storage in the CW – nonstock items (Egbelu et al., 1998; Aptel & Pourjalali, 2001; Landry & Beaulieu, 2010b). This requires the establishment of collaborative relationships in the supply chain, by emerging collaboration concepts adopted in the healthcare industry (Carvalho, 2010). In order to integrate the internal and external supply chains, the practice of stockless materials management programs began to appear in this sector (Rivard & Royer, 2002). In this practice, distributors are responsible for the supply and distribution of material directly to the patient care units eliminating one connection in the supply chain, the hospital CW (Landry & Philippe, 2002). In the definition of Vendor Management Inventory (VMI), the stock management becomes the supplier's responsibility. The supplier controls the amount of inventory in the hospital (costumer) according to information on inventory consumption and has replenishment stock responsibility (Carvalho, 2010). In cases where customers have more power, the stock is at the customer facilities and on consignment. Hospital Professor Doutor Fernando Fonseca (Neves, 2009) and the Centro Hospitalar de Lisboa Norte (Carpenter & Viegas, 2012) have implemented VMI. The consignment is defined according to Lee (2008) as the process in which the supplier puts the material at the customer's facilities, without receiving payment until the material is consumed. In an audit carried out in Portugal by Carpinteiro and Viegas (2012) all audited hospitals, a total of six, acquire medical supplies on a consignment basis. Some of these cases are the Hospital of Vila Nova de Gaia/Espinho (Teixeira, 2012) and the Hospital of Santa Maria (Nabais, 2009).

The supply of material to patient care units plays an important part of the hospital's internal supply chain. Thus, through an adequate supply of material, an organization can achieve lower operating costs by reducing investment in inventory handling and improving service levels, providing better healthcare to patients (Perrin, 1994). The distribution process of materials to patient care units can be accomplished in several ways. A number of replenishment systems were adopted by the hospitals, over time and across the world, which are reported in the literature. The most traditional replenishment system is the requisition system. In this system the healthcare professional controls the process of deciding when and how much to order (Wolper, 2004). Despite being simple and with low investment, it generates extra requisitions, unnecessary high stock levels or unacceptable low stock levels, and takes up too much of healthcare professionals time (Perrin, 1994; Osorio, 2009). For many years the trend has been changing the traditional system for the exchange carts system, relieving healthcare professionals of the replenishment task (Perrin, 1994). In this system, the material is placed on a movable cart positioned in the storage unit. Another identical cart with exactly the same stock is prepared. The cart trading is done according to a predetermined schedule by the second cart (fully stocked). The first cart returns to the CW to be replenished and later will replace the cart that is in the storage unit (Landry & Beaulieu, 2010a). According to Wolper (2004), this system requires a large investment in carts and the inventory is duplicated. At par level system when the inventory quantities are below the established level, the material is restored to the maximum level. Some hospitals use a min-max variation of this system (Wolper, 2004; Landry & Beaulieu, 2010a). The par level system is implemented at Vila Nova de Gaia/Espinho's hospital, supported by IT (Teixeira, 2012). In the two-bin system, each item is stored and distributed over two containers. The material is only removed from one container. When the first container is empty the healthcare professional uses the second container. Before that, the need for replacement is identified. This can be done in several ways, from identifying the withdrawing label of the product and adding it to a specific location, sending the empty container to the CW to be restocked, etc. The Kanban method is used to set up the signal (Landry & Beaulieu, 2010b; Leone & Rahn, 2010). Later new practices associated to the two-bin system arise, like RFID-enable two-bin system and e-kanban (supported by an IT system) (Landry & Philippe, 2002, 2004; Bendavid, 2010). This system is implemented in hospitals in various countries like USA, France, Japan, Portugal, among others (Landry & Beaulieu, 2010a). Portugal Pulido Valente's Hospital (Lima, 2007) and Santo António's Hospital already adopt this system (Fontes, 2005). Other systems were introduced, like automated storage cabinets, which were developed as a way to link patient consumptions with supplies, and weight control bin method, where the replenishment is triggered when the bin reaches the preset weight for each product type (Landry & Beaulieu, 2010b). According to Bendavid et al. (2010) U.S. hospitals are adopting the RFID-based cabinets for high value products. Each replenishment system has its own advantages, disadvantages and ways of operation. However according to literature review conducted under this project, healthcare organizations have been adopting only a single replenishment system for the distribution of medical supplies.

3. Medical supply distribution in Braga's Hospital

HB is integrated in the National Health Service's network and provides healthcare to about 1.2 million people in the districts of Braga and Viana do Castelo (Braga Hospital, 2015, paragraph 3). The hospital's internal chain includes 1 CW and 69 storage units. The CW is a large warehouse, with a large number of material references (higher than 1600) and with the main function of supplying all storage units. Each storage unit works as small-scale storage points, located within the patient care units which they supply, allowing healthcare professionals to gain quick access to the materials. On HB the medical supplies are classified under three categories according with the supply strategy:

- Storage material: material with high turnover, usually common to several clinical departments, and stored in the hospital's CW, hence the name of storage material. This type of material can be divided into two sub-categories:
  - Material without registration: covers all materials where it is not possible to register it to patient's treatment, such as caps, gowns, garbage bags, and more.
  - Material with treatment's consumption record: the materials used during the treatment are recorded and associated with the procedure and the patient, such as syringes, needles, etc.
- Specific material: specific material for specific clinical departments, usually characterized by being low movers material and in most cases with a high cost (e.g. catheters). This material is delivered directly to the patient care unit without the need for storage in CW (cross –docking).
- Material on consignment: The management of consigned material is the supplier's responsibility. The hospital only has the responsibility to pay after using it. This type of material is not stored in the CW and is delivered directly to the patient care unit, getting stored in the storage units (cross–docking). The cost of such material is generally very high and such materials are only for specific specialties (e.g. implants). Cross-docking operations represent 53% of all hospital's CW receptions.

The replacement of medical supplies in the patient care units is performed in three replenishment logistic systems according to the type of material.

3.1. Replenishment systems

3.1.1. Two-bin system. The replacement of the storage material, without or with consumption registration, is performed through the two-bin system with Kanban cards. Kanban is a card that is associated with two boxes of material and consists of a label with the product barcode. The barcode information allows you to obtain the material's specifications and the amount to be restored.

Healthcare professionals start by taking material from the first box. When the last material unit drive ends (leaving an empty box) the Kanban card is removed and placed in a separate box (unread cards box), where the cards are to be scanned, and then they start to take the material from the second box. These cards of empty boxes are read by PDA (Personal Digital Assistant) from the storekeeper and a request is made to the CW to pick up the requested products. The Kanban cards that are in the "unread cards box" are moved to "read cards box". The request of material is triggered by the material consumptions - Kanban signal. The replacement of requested materials is held on the same day. Once the materials are replaced, cards are placed again, from the "read cards box" onto the respective product box. At the hospital there is a weekly plan already established that sets the number of times per week that the storekeeper has to scan the needs for each patient care unit. In some cases, the material requested by the patient care unit may be in stock-out in CW. As a result there is no replacement of these materials, through the cards from "read cards box" to the "unread cards box" so that the next scanning needs to be read back again. This does not allow healthcare professionals to understand whether the material is in stock-out in CW or if that material has already been delivered and consumed. The reordering point in a two-bin system was calculated considering an average consumption in a period between replacements, more another day - safety stock. This replenishment model combines features of continuous and periodic review models since, every review period (fixed-interval), if the quantities in stock are below the minimum (kanban signal), a fixed order is placed (one box) to achieve the maximum level (two boxes).

3.1.2. Requisition system. The replenishment model used for the specific material is the traditional or requisition's system. In this model, healthcare professionals are responsible for monitoring and controlling material inventory levels and to place orders to Purchasing department, when they consider it necessary. According to healthcare professionals this task takes, on average, about 1 of the 12 working day hours. This time was not timed but indicated by healthcare professionals based on their experience. In this type of materials only a maximum level is defined. These levels are defined by clinical personnel together with the Logistics department.

When the specific material is delivered to the CW, an input and an output of the material is given on the IT system, which means that the material is marked as consumed even before reaching and being available for use at the specific patient care unit. From this moment on, there is no more control or tracking information of this material.

3.1.3. Consignment. In consignment, the supplier is responsible for stock management. After the consumption of this type of material healthcare professionals just need to register on a specific sheet the material that was used and send it to the Procurement department where an input and an output is given at the same time of the correspondent material on IT system ,before make the purchase order. In this replenishment model, all consignment materials have a maximum level. This level is defined by Logistics department along with suppliers and with the clinical personnel. The replenishment policy is based on the following assumption: whenever there is consumption of a unit, the supplier restores a unit of this material.

The control of the material is limited. Although the supplier has the responsibility of stock management there is no control of the material that is placed in the hospital by the supplier or the material quantities available in storage units. This is because the HB has neither the function to check or to give input and/or output of the material delivered by the supplier. In addition the logistic process, since the consumption until the registration of consumed material, is set in different locations and times, by more than one agent, and by the use of paper.

Table 1 summarizes the main problems encountered in the existing replenishment systems. Based on this analysis there is a need to get involved through the development of this project in order to remove the limitations found.

Table 1. Identified problems on the existing replenishment models.

Replenishment system	Problems
Two-bin system	- Difficulties on matching the patient's consumption records; - Different times between the need to and the request of a material replacement; - Difficulty in identifying if the material has been consumed or is in stock out in CW.
Requisition system	- Healthcare professionals are responsible for stock management ; - Nursing deviation from their main tasks; - Low inventory and consumption control.
Consignment system	- Low control of the material delivered in hospital facilities by suppliers; - Consumption to record process is bureaucratic and happens in different places.

4. Proposed solution

After identifying and analyzing the existing problems in the replenishment systems, the adopted solution was to use AW which are an extension of CW, acting as supermarkets, supported by IT system. The AW model is based on: par level system for specific materials and storage materials with treatment's consumption record; two-bin system for storage materials without registration; and consignment model for consigned materials. The operation of all models is supported by the IT system which allows a connection of the records of product consumption and the stock levels. These three systems will operate at the same time working in parallel on the same patient care units. This approach is innovative because in other hospitals where AW concept is implemented, it is only based on one replenishment system. These hospitals typically use or par level system or two-bin system.

This solution was initially implemented in Angiography unit due its low activity and reduced levels of medical supplies and consequently in COR which is a service with high activity levels, with large amounts of material of different specific clinical departments (representing 50% of total consumption of HB materials). Of the total stock value of these two patient care units and the CW, Angiography represents on average 2,85%, COR represents 39,15% and CW represents 58%.

4.1. Par level system implementation.

Through the implementation of AW concept with par level system, all materials stored in AW have associated a maximum and a minimum level on the IT system. While the maximum level is the maximum possible amount to own, the minimum level corresponds to the reorder point. This replenishment system is the engine of min-max system. If the quantity is below the minimum level, an order is triggered with a fixed amount (difference between maximum and minimum level). In this system, the traceability of the products is high allowing the user to know in which AW are certain products, at a given instant. Material consumption records are preformed on time and the levels are instantly updated on IT system. Material registration is essential, matching real stock on the ward with the stock on the IT system, since the absence of it causes a mismatch between them.

4.1.1. Treatment's consumption record material. The consumed materials are no longer being replaced by the two-bin system and they are replaced through the AW concept using par level system. In this new model the maximum levels are the same but the minimum levels are decreased to reduce inventory levels due to the fact that the minimum level does not need to be half of it (one box in the two-bin system). This reduction is possible since the patient care unit supply is performed on the same day. This system provides a proactive replacement with more automated and timely replenishment orders (at the time of consumption), without the need for a storekeeper to go through all the clinical departments to read the material needed. These materials represent 46,1% of all materials value.

4.1.2. Specific material. With the AW concept implementation with par level system, the healthcare professional fails to monitor the stock levels of specific materials, passing the responsibility to Procurement department to analyze material's levels. The products that have less than the minimum level are followed by a purchase order to the respective supplier.

The supplier's material is received in the CW's receiving area, recording all items received on the IT system in the respective AW, by reading the bar code with reference, batch and expiry date. When there is consumption, the healthcare professionals register all the material's data into the IT system. The minimum and maximum levels were defined based on the last 12 months purchases activity since there were no records of consumption, to cover the service for a month. These levels are justified for being in an early stage of the project, therefore it was important to have a bigger safety margin. Control of this type of material increased considerably. Previously, the only control that existed was when an input and an output was given on the IT system, after the material was received in the CW. It is now possible to know when the material arrives, when it is consumed and when there is need for replacement in a patient care unit, the level of quantities, among others. The specific material in AW replaced by par level system, represents approximately 7,1% of all materials value.

4.2. Consignment

The request of consigned material, unlike the previous process, is triggered by the registration of consumption in patients care units at the time of their use by the same agent (healthcare professionals). The replenishment system is exactly the same as practiced until now. Whenever there is material consumption, the unit is replaced. After receiving the material in CW, the entry record is given on IT system consignment model, by reading the barcode with the reference, batch and expiry date.

Every day an email is sent to the suppliers specifying which material has been consumed. This email is sent even if there is no consumption of any product to avoid possible confusion next to the supplier, because if it is only sent when there is consumption, any problem sending emails, it could easily be interpreted by the supplier as lack of consumption even if it had existed.

In this new model the responsibility of material management remains with the suppliers. However, there is more control of products that are delivered by suppliers to the hospital facilities. In addition the IT system does not allow the suppliers to deliver material that does not belong to the material list agreed between the hospital and the suppliers. The consignment material in AW represents 27,7% of all materials value.

4.3. Two-bin system

Replenishment process of storage material without registration continues to be held by the two-bin system with Kanban cards. The two-bin system is already established and implemented in most hospital patient care units and it is suitable for use with this type of material. Therefore it continues to be implemented in AW of the patient care units. The Logistics department already has a solution planned for giving feedback information to the healthcare professionals in case of sotck-outs in CW, by placing a paper saying "Stock out" in the container where the material should be placed. As a result, healthcare professionals know what materials are in stock out in CW. The material in AW replaced by two-bin system represents 4% of all materials value. The rest 15% of all materials value are related to the non-resident consignment material (material that are not in the hospital facilities, only used in specific interventions and that arrives on the same day of the respective intervention).

5. Result's analysis

Regarding material with treatment's consumption record, the material replenishment is based on consumption records, allowing firstly, detect the need for replacement faster (at the time of use) and secondly, make the orders automatic (picking request is made by IT system to the CW).

With regard to the specific material, the great advantage of this implementation is that the management of those materials is no longer the responsibility of healthcare professionals and becomes the Logistics' responsibility, freeing about 8% of the healthcare professionals working time to other activities, thus improving healthcare quality. There is also greater material control with access to inventory levels in real time and with the information of consumptions records in patient care units the replacement becomes then triggered on the basis of consumption.

At consignment, in addition to the improved material control delivered to the hospital, it was also possible to cut down on paperwork during the logistics process, since previously consumption recording was made on a sheet in patient care unit, before being sent to Procurement department to record the consumption on IT system.

In addition to the benefits previously mentioned, it was also possible to obtain other advantages. One of them is related to the expiration time control. It is possible to alert healthcare professionals to consume first the materials with shorter expiration time. Another important aspect was changing the management of specific materials to consignment basis. These materials have a very high cost and by managing on a consignment basis it was possible to reduce stock capital costs. However some materials are not yet on consignment basis since they are already in stock and paid for. Thus these materials only swap after being consumed.

With the development of this project these three replenishment systems coexist in the hospital supply chain, running in parallel according to the type of medical supplies (Figure 1):

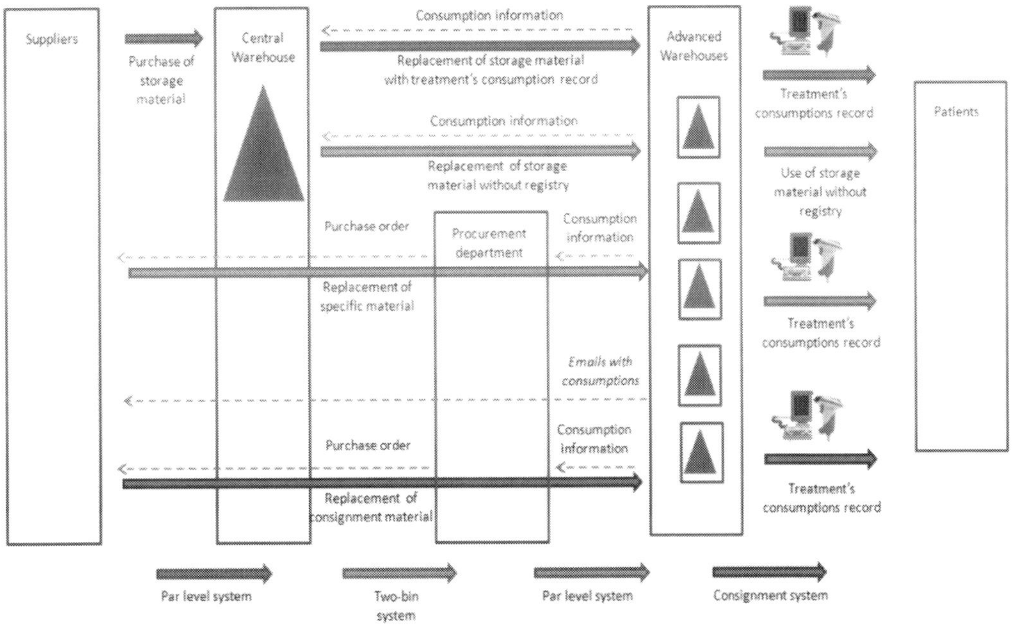

Figure 1. Hospital Supply Chain with AW concept in use.

Table 2 shows the stock units and references before and after the AW concept in Angiography and COR is implemented. With this system we were able to obtain an average holding stock value reduction of about 28.2% in Angiography. The increase number of references in Angiography may be justified by reduced stock outs, i.e., due to the fact that they weren´t available in this store unit before this implementation.

Table 2. Stock averages before and after the implementation of AW in Angiography and COR.

Patient care unit	Date	Stock units	References
Angiography	Before AW implementation	511	32
	After AW implementation	634	45
COR	Before AW implementation	28573	964
	After AW implementation	32152	1017

With regards to COR, the average value of holding stock was reduced by approximately 4.6% with the implementation of AW, even with the increasing number of units and number of references, like in Angiography (Table 2). This value will be further reduced with the material changeover from specific to consignment.

The COR's material service level increased considerably. This calculation is based on the number of requests that were satisfied by the CW into the COR. The service level went from about 91% to about 97% with the AW concept implementation (Table 3). The high level of service requests is partly justified by the high inventory levels in AW and due to the fact that it started recently.

Table 3. COR service quality before and after the implementation of AW concept.

	Total requested units	Total satisfied requested units	% satisfied requested units
Before AW implementation	813952	736879	90,79%
After AW implementation	621697	605728	97.43%

6. Conclusion

Due to rising material costs and increased competitiveness in this sector, healthcare organizations felt the need to develop an economically sustainable material management that at the same time could provide quality healthcare services to patients. It is essential that the medical supplies are always available for patient's treatment and that healthcare services take place normally. Logistics is then a key role to support inventory management, supply and distribution policies so that the service level is not affected.

Before this project was in place, the patient care unit's supply process still used old-fashion methods with the participation of healthcare professionals, and with inefficient management and control of stocks by Logistics department, resulting in high stock levels of certain materials and disruptions with other materials. The impact of the AW concept in the hospital's patient care units is huge since it reduces inventory levels, reduces the associated costs with material management, increases service level, and improves control over stocks and satisfaction level of healthcare professionals and patients.

Although this implementation has advantages it also presents some challenges: In certain materials there is a mismatch between the real stock and the IT stock, affecting the material's replacement. This gap is due to the fact the records of materials consumed are not always made by healthcare professionals, the fact that other clinical departments withdraw material from Angiography and COR departments, and in COR, there is stock in the surgery rooms.

With the development of this project, we intend to give relevance to the complexity of the supply chain of healthcare organizations, as well as the large number and variety of materials that these organizations deal with daily, and demonstrate that the integrated operation of several replenishment systems, according to the category of the material, may result in benefits for these organizations. Besides that, it is essential to note that IT plays an important role to improve the performance of the replenishment process.

Acknowledgements

This study had the financial support of FCT-Fundação para a Ciência e Tecnologia of Portugal under the project PEst- UID/CEC/00319/2013.

References

Aptel, O. & Pourjalali, H., 2001. Improving activities and decreasing costs of logistics in hospitals: a comparison of U.S. and French hospitals. *The International Journal of Accounting,* 36(1), pp. 65-90.

Bendavid, Y., Boeck, H. & Richard Philipe, R., 2010. Redesigning the replenishment process of medical supplies in hospitals with RFID. *Business Process Management Journal,* 16(6), pp. 991-1013.

Callender, C. & Grasman, S. E., 2010. Barriers and best practices for material management in the healthcare sector. *Engineering Management Journal,* 22(4), pp. 11-19.

Carpinteiro, J. A. & Viegas, M. I., 2012. *Auditoria ao aprovisionamento das unidades hospitalares do SNS,* Lisbon: Tribunal de Contas.

Carvalho, C., 2010. *Logística e Gestão da Cadeia de Abastecimento.* Lisbon: Edições Sílabo.

Carvalho, C. & Ramos, T., 2010. *Logística na Saúde.* 2nd ed. Lisbon: Edições Sílabo.

de Vries, J., 2011. The shaping of inventory systems in health services: A stakeholder analysis. *International Journal of Production Economics,* 133(1), pp. 60-69.

Egbelu, P. J., Harmonosky, C. M., Ventura, J. A., O'Brien, W. E. & Sommer, H. J., 1998. Cost analysis of hospital material management systems. *Journal of the Society for Health Systems,* 5(4), pp. 1- 10.

Fontes, N., 2005. *Hospital Logistics System.* Faculdade de Engenharia da Universidade do Porto.

Hospital de Braga, 2015. O Hospital de Braga. [Online]
Available at: www.hospitaldebraga.pt
[Accessed 06 February 2015]

Kelle, P., Woosley, J. & Schneider, H., 2012. Pharmaceutical supply chain specifics and inventory solutions for a hospital case. *Operations Research for Health Care*, 1(2-3), pp. 54-63.

Landry, S. & Beaulieu, M., 2010a. Achieving lean healthcare by combining the two bin-kanban replenishment system with RFID technology. *International Journal of Health Management and Information*, 1(1), pp. 85-98.

Landry, S. & Beaulieu, M., 2010b. *The Challenges of Hospital Supply Chain Management, from Central Stores to Nursing Units*. In: Brian, T. D. (ed). Handbook of Healthcare Operations Management - Methods and Applications. New York: Springer Science & Business, pp. 465-482.

Landry, S. & Philippe, R., 2002. *4U2C, or How logistics can service healthcare*. Quebec: Ecole des Hautes Commerciales.

Landry, S. & Philippe, R., 2004. How logistics can service healthcare. *Supply Chain Forum*, 5(2), pp. 24-30.

Lee, W., 2008. Managing level of consigned inventory with buyers warehouse capacity constrained. *Production Planning & Control: The Management of Operations*, 19(7), pp. 677-685.

Lee, S. M., Lee, D. & Schniederjans, M. J., 2011. Supply chain innovation and organizational performance in the healthcare industry. *International Journal of Operations & Production Management*, 31(11), pp. 1193-1214.

Leone, G. & Rahn, R. D., 2010. *Lean in the OR*. Boulder: Flow Publishing.

Lima, P., 2007. *O Duplo Lote*. Lisbon: Centro Hospitalar de Lisboa Norte, EPE.

McKone-Sweet, K. E., 2005. The ailing healthcare supply chain: a prescription for change. *Journal of Supply Chain Management*, 41(1), pp. 4-17.

Nabais, J., 2009. *Inventory management for the Health Sector ABC analysis approach*. Faculdade de Economia da Universidade Nova de Lisboa.

Neves, R., 2009. *Projecto de melhoria da logística hospitalar do hospital Infante D. Pedro*. Universidade de Aveiro.

Osório, P., 2009. *Impacto da implementação de armazéns avançados em serviços do Centro Hospitalar Vila Nova de Gaia-Espinho, EPE*. Universidade Nova de Lisboa.

Ozcan, Y., 2009. Quantitative *Methods in Health Care Management: Techniques and Applications*. 2nd ed. San Francisco: Wiley.

Perrin, R. A., 1994. Exchange cart and par level supply distribution systems: form follows function. *Hospital materiel management quarterly*, 15(3), pp. 63-76.

Pinto, J., 2008. *Kaizen nas unidades hospitalares criar valor eliminando desperdício*. Faculdade de Engenharia da Universidade do Porto.

Rivard-Royer, H., Landry, S. & Beaulieu, M., 2002. Hybrid stockless a case study lessons for healthcare supply chain integration. *International Journal of Operation & Production Management*, 22(4), pp. 412-424.

Rossetti, M., 2008. *The Inventory Management Issues in Health Care Supply Chains*. University of Arkansas.

Schneller, E. & Smeltzer, L., 2006. *Strategic Management of the Health Care Supply Chain*. San Francisco: Jossey-Bass.

Teixeira, D., 2012. *Armazém avançado em contexto hospitalar: método de reposição por níveis*. Universidade de Aveiro.

Wolper, L., 2004. *Health Care Administration: Planning, Implementing, and Managing Organized Delivery Systems*. Sudbury: Jones & Bartlett Publishers.

Journal of Physics: Conference Series **616** (2015) 012006 doi:10.1088/1742-6596/616/1/012006

Planning Nurses in Maternity Care: a Stochastic Assignment Problem

Frank Phillipson

TNO, POBox 96800, 2509 JE The Hague, The Netherlands

E-mail: frank.phillipson@tno.nl

Abstract. With 23 percent of all births taking place at home, The Netherlands have the highest rate of home births in the world. Also if the birth did not take place at home, it is not unusual for the mother and child to be out of hospital in a few hours after the baby was born. The explanation for both is the very well organised maternity care system. However, getting the right maternity care nurse available on time introduces a complex planning issue that can be recognized as a Stochastic Assignment Problem. In this paper an expert rule based approach is combined with scenario analysis to support the planner of the maternity care agency in his work.

1. Introduction

The main view in The Netherlands is that childbirth is not a medical condition and pregnant women should not be treated as patients. As a result pain relief is not encouraged and home births are encouraged.

The role of the doctor or gynaecologist in a normal pregnancy in The Netherlands is minor and in most cases they are not involved at all. The central person in the pregnancy is the midwife. With 23 percent of all births taking place at home (decreased from 38% in 1990), The Netherlands boasts the highest rate of home births in the world. In Britain home births account for 3 percent of total births, in Belgium and Germany this figure is closer to 1 percent. Similarly 99 percent of births in the United States take place in a hospital [1].

Most probably the main merit of giving birth in The Netherlands lies in the post-natal care [2]. Also if the birth did not take place at home, it is not unusual for the mother and child to be out of hospital in hours after the baby is born. There is a logical explanation for the short post natal care in Dutch hospitals: the maternity care system. The Dutch maternity care nurse is the envy of many women outside of the Netherlands. This type of maternity care is pretty much unique. For at least a week after the birth professional help is on hand. During a home birth the maternity care nurse supports the midwife and after a hospital birth the maternity care nurse is on your doorstep within hours of leaving hospital.

Maternity care duties range from care for the new mother and infant, light household duties, guidance on breast feeding and baby care and looking after other family members (such as other children). Every pregnant woman in the Netherlands has the right to post natal maternity care. The exact details of what a particular maternity nurse will deal with, as well as the frequency and length of her visits, will be determined before birth. This prenatal house-call takes place at around the 7th or 8th month and allows both parties to discuss the needs and expectations.

Therefore a pregnant woman has to register with a maternity care agency (MCA), preferably before the 12th week of pregnancy, to be sure to get this unique care.

When labour starts and the birth is planned at home, both the midwife and the MCA are warned. When the birth is planned at the hospital or moved to the hospital just before or during the childbirth, the MCA is warned when the mother and child leave the hospital. When the MCA is warned they will send a nurse to the home. Normally several nurses are stand-by or on duty. Here the question arises which nurse should be sent to the specific house.

2. Challenge

For the MCA it is important to provide adequate care during an adequate number of hours, and at the right time. The MCA needs an expert system providing an admissible solution in due time by assigning the right nurse to each upcoming demand, weighting the effect this assignment has on decisions in the future, under uncertainty. Nowadays, if there is a demand for a nurse, the planner of the agency should take into account a large number of variables in the decision of which of the available nurses has to be assigned, like:

- skills of the nurse should comply to the demand;
- the nurse should be able to work in a certain house, where people do or do not smoke, with or without other kids or pets, where are stairs etc.;
- the workload should be evenly distributed but nurses should work a number of hours as close to their contract as possible;
- meet health and safety standards;
- comply to work schedule agreements;
- perform high on quality standards like reaction time and number of changes in nurse.

Although most experienced planners succeed in this assignment in most cases, structural incorporation of the future effects of a decision is hard for a human. Computer assistance is welcome here. This expert system should work based on mathematical algorithms and use the already digitally available but continuously changing data on demand side (duration, size, location) and supply side (contract size, residence, utilization, deployment constraints, schedules). Such expert systems exist in other sectors (regular nurses, logistics, telecommunications, energy) but is not yet available in maternity care industry.

The planning methodology must find a balance between productivity of workers and the quality of care received. An important indicator of the quality are the number of hours received and the number of changes of nurses. More than one change (two different nurses) is not desirable. In this paper an approach is proposed for this problem.

3. Problem description

The key problem here is that a nurse should be assigned as soon as an incoming demand arrives. The planner has a number of possibilities here: the planner can

- Use one of the nurses available due to a scheduling mechanism; nurses have had some free days and are now available for a new task (on the 'bench');
- Replace an existing assignment by taking the nurse from another task to the new task and send a nurse from the 'bench' to the free coming task;
- Use an external nurse, on call-basis, with no contract based on a fixed number of hours.

Note that the nurses 'on the bench' are already paid and the external nurses have to be paid based on the hours they make. Their cost per hour will be higher. This assignment should be such that it optimizes the performance indicators of the agency, like profit, number of delivered hours, number of nurse changes etc. It should also comply with constraints that consider skills of the nurse and allergies.

As an example take a new task for which two nurses are available on the 'bench'. One of the nurses (1) is willing to serve in a house that has snakes as pets and needs hours to fulfil her contract. The other nurse (2) does not like snakes and is already high in her hours that month. The planner knows that there is a client, just passed her due date, that has snakes. What does he do, save nurse 1 for the client with a snake, creating the probability that she does not fulfil the hours in her contract or not, enlarging the probability on a nurse change (one of the performance indicators)?

From a mathematical point of view the problem can be seen as a special case of the sequential stochastic assignment problem as discussed in [3, 4]. This problem addresses the assignment of n IID sequentially-arriving tasks to n available workers (resources). Choosing a certain worker to perform the arriving task renders him unavailable for future assignments, with an expected reward associated with this assignment. The objective is to assign the n workers to n arriving tasks so as to maximize the expected total reward obtained from pairing workers with tasks. The difficulty in making the assignment decisions is that the assignments are performed in real-time. Once a task arrives and its value is revealed, it must be paired instantly with one of the available workers, without knowing any of the future task values. Although the main idea is the same in the nurse assignment problem, assign unknown arriving tasks to one of the available workers, several characteristics of the nurse assignment problem are different:

- The nurses come back in the system after their job finished.
- The nurse can stop the current job, start a different one, having another nurse replacing her in the first job.
- There are more tasks than nurses.
- A nurse can handle more than one job together on the same day (e.g., two duties of 3 hours).
- The nature of the future jobs are known, only the arrival time is stochastic.
- The problem is multi-objective.
- etcetera.

This makes the problem special and not discussed earlier.

4. Approach

The main idea of the approach presented here to assist the planner on this assignment decision is that of scenario analysis: determine for each possible assignment the expected effect on the performance indicators. For this, for every possible assignment a simulation is performed, where a huge number of possible realisations of the future (all the births) are generated. For each of these realisations a planning is created; which nurse is doing which task. The method that is used for this is a greedy rule based forward assignment model [5]. Next a score has to be assigned to each planning.

In Algorithm 1 the main procedure is described. For the assignment of a given task t, the arriving demand for a nurse, for all the available nurses that have the minimum skills for this task a set of simulation is performed. The available nurses are the nurses on the 'bench', all active nurses on another (interruptible) job and the set of external nurses. In other words, first only the nurses on leave (rest period, holiday) are not considered. Next, if a certain skill is demanded, the nurse have to fulfil this request. This can be some skills on breastfeeding, special care or the willingness to serve in a house with smokers, certain animals, other kids etc.

For each possible assignment a set of replications of the simulation ($NumberReplications$) is performed. In each simulation a realization is determined where the dates of all the delivery dates

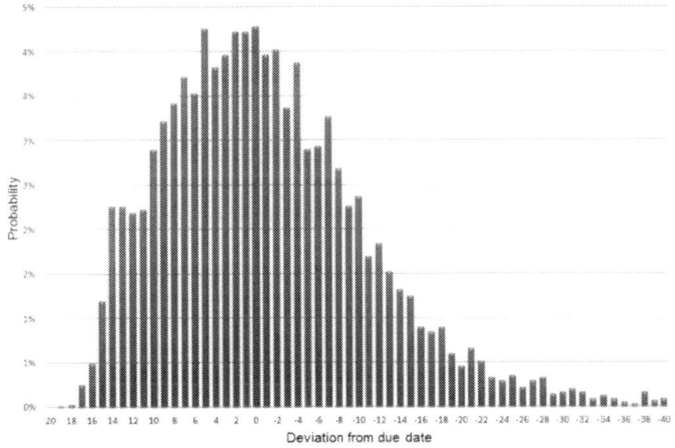

Figure 1. Probability deviation from due date

are fixed following the probability distribution, see the example in Fig. 1. For each realisation an assignment is calculated. This assignment is scored on the performance indicators as defined by the MCA.

Algorithm 1 Main Approach

Require:
Ensure: Assignment of task t
for $i \in$ AvailableNurses **do**
  if Nurse i can perform task t **then**
    $j = 1$
    $N = NumberReplications$
    while $j < N$ **do**
      Create dates for all deliveries resulting in tasks $t_1, ..., t_n$
      P_j=CreateGreedyPlanning($t_1, \ldots, t_n$)
      Calculate score $S_j(P_j)$
      j=j+1
    end while
    Score of nurse i: $TS_i = \frac{1}{N} \sum_{j=1}^{N} S_j$
  end if
  Assign task t to nurse $i = \text{argmin}_k\, TS(k)$
end for

How to create the planning is explained in Algorithm 2. Here a rule based assignment is chosen to get an acceptable calculation time. The exact quality of the solution is not the most important issue, where the planning should only reflect a picture of the effects of an initial assignment.

Based on expert opinion of the current planners of the agency three properties are translated to the rule base:

(i) Accomplishment: which part of the task (in hours) can be performed by nurse i. For example, if a nurse just started working and can work 50 hours until the next rest period and the task is 80 hours, the accomplishment equals $50/80 * 100\% = 62.5\%$.

(ii) Utilization: which part of the contract hours has not be assigned. For example, if a nurse works 140 hours in a period (4 week, month) and she has already worked 40 hours, the utilization equals $100/140 * 100\% = 71.5\%$

(iii) Distance: what is the distance of the nurse to the address of the task. Zero km equals a score of 100%, some maximum distance (e.g. 30 km) equals a score of 0%.

Here all remaining tasks, as drawn in the realization, have to be assigned. For each task, sorted in step 1 first on starting day and within the same starting day on the length of the task (longest first), again all available nurses are considered for assignment. Available has the same meaning as before, only now also rest periods are added. Given legal health and safety standards a rest period of two or three days is placed after a working period of a certain number of days and working hours.

For an available nurse the three properties, Accomplishment, Utilization and Distance, are calculated and added together to form a score. The weight (w_i) of this summation are adjustable. There are also minimum values on each property. This is done in step 9. Under that minimum the total score should be zero. This can also be realized by making the total score a product of the three value instead of a sum.

Now the nurse with the highest total score is assigned to the task.

Algorithm 2 CreateGreedyPlanning

1: Sort $\{t_i\}_{i=1}^{n}$
2: **for** j=1 to n **do**
3:   **for** $j \in availablenurses$ **do**
4:     Calculate $s_1 = accomplishment(t_i, n_j)$
5:     Calculate $s_2 = utilization(t_i, n_j)$
6:     Calculate $s_3 = distance(t_i, n_j)$
7:     $score[j] := w_1 * s_1 + w_2 * s_2 + w_3 * s_3$
8:     **for all** k=1 to 3 **do**
9:       **if** $s_k < min_k$ **then**
10:         $score[j] := 0$
11:       **end if**
12:     **end for**
13:   **end for**
14:   Assign task i to nurse $j = \text{argmax}_k\, score(k)$
15: **end for**

5. Conclusions

In this paper an approach was presented to support the planner of a MCA in assigning the right nurse to an incoming demand. Right means here that it optimizes the performance indicators in the long run. Especially this long run makes it hard for a human planner. For this, a scenario analysis based expert system was proposed where various assignment decisions are elaborated on a huge number of possible realisations of the future (the births). The first results of using the tool in practice seemed promising, for an agency with 260 expected deliveries and 180 nurses, the tool gives a proposal within 2 minutes. However, due to inaccessibility of data in other, related, systems, the fully automatic version could not be tested for a long period, this is future work. Also refinements of the expert rules and the effect of them on the outcome are open for future work.

References

[1] Chervenak F A, McCullough L B, Brent R L, Levene M I and Arabin B 2013 *American journal of obstetrics and gynecology* **208** 31–38
[2] Maternity matters: What to expect in the Netherlands URL `www.expatica.com`
[3] Derman C, Lieberman G J and Ross S M 1972 *Management Science* **18** 349–355

[4] Baharian Khoshkhou G 2014 *Stochastic sequential assignment problem* Ph.D. thesis University of Illinois at Urbana-Champaign
[5] Azibi R and Vanderpooten D 2002 *European Journal of Operational Research* **138** 274–293

Journal of Physics: Conference Series 616 (2015) 012007 doi:10.1088/1742-6596/616/1/012007

Introducing health gains in location-allocation models: A stochastic model for planning the delivery of long-term care

T Cardoso[1,3], M D Oliveira[1], A Barbosa-Póvoa[1] and S Nickel[2]

[1]Centre for Management Studies, Instituto Superior Técnico, Universidade de Lisboa, Lisbon, Portugal

[2]Institute of Operations Research, Karlsruhe Institute of Technology, Karlsruhe, Germany

[3]Corresponding author: teresacardoso@tecnico.ulisboa.pt

Abstract. Although the maximization of health is a key objective in health care systems, location-allocation literature has not yet considered this dimension. This study proposes a multi-objective stochastic mathematical programming approach to support the planning of a multi-service network of long-term care (LTC), both in terms of services location and capacity planning. This approach is based on a mixed integer linear programming model with two objectives – the maximization of expected health gains and the minimization of expected costs – with satisficing levels in several dimensions of equity – namely, equity of access, equity of utilization, socioeconomic equity and geographical equity – being imposed as constraints. The augmented ε-constraint method is used to explore the trade-off between these conflicting objectives, with uncertainty in the demand and delivery of care being accounted for. The model is applied to analyze the (re)organization of the LTC network currently operating in the Great Lisbon region in Portugal for the 2014-2016 period. Results show that extending the network of LTC is a cost-effective investment.

1. Introduction

Long-term care (LTC) aims at improving the quality of life of individuals who are dependent on help with basic activities of daily living due to chronic illness and/or disability [1]. LTC includes the organization of both nonmedical and medical services, with these varying across countries [1]. In particular, different services can be provided (with services ranging from short-term to longer institutionalizations, home-based and ambulatory services) and different entities can be responsible for providing care (e.g., family, public, private and not-for-profit entities).

The European LTC sector has been facing several challenges, as there has been an increasing demand for LTC, mainly due to the ageing phenomenon and to the increasing prevalence of chronic diseases [2]. Furthermore, the current supply of LTC is clearly not enough to meet this growing demand [2]. Within this context, planning the delivery of LTC is a top policy priority in many European countries. Particularly, when facing a strong pressure to decrease and control public spending in health care, countries based on a National Health Service (NHS) structure need to plan networks of LTC that minimize costs and account for other top level system objectives, such as the attainment of health gains and of equity. To help such planning, decision support tools are required.

Mathematical programming models have been broadly used in the health care planning literature [3], in particular for planning the location-allocation of services. Nevertheless, there is a lack of studies comprehensively modeling key features of health systems that are critical for proper planning in real settings. In particular, although a vast literature in the area of health care planning relies on the use of single objective mathematical programming models [4], only recently models with multiple objectives have been proposed in this area (examples can be found in [5-7]). Within the existing models, equity [8], costs [7] and health benefits [9] related objectives have been accounted for, but to the authors' knowledge, no study has addressed the joint attainment of these different dimensions. Also, a vast health care planning literature relies on equity or cost related objectives (such as [7] [8,10]), whereas health benefits has seldom been used (examples can be found in [9,11]). Also, few research has addressed uncertainty in an integrated way, for instance using stochastic models [12]. To the authors' knowledge, the only stochastic models proposed for strategic and tactical planning of health care services can be found in [6,7,13]. Furthermore, few studies have modelled multi-service systems, and decomposed the planning horizon into periods [7,14]. Additionally, very few studies were dedicated to the LTC sector [15], with no LTC study comprehensively addressing all the features that are relevant for health systems.

This paper develops a multi-objective two-stage stochastic mixed integer linear programming (MILP) model to support location-allocation decisions in the LTC sector in the context of a NHS-based country. The proposed model assists LTC planners on how to (re)organize a multi-service network of LTC (including a wide range of institutional, home-based and ambulatory care services), providing information related to capacity planning and location selection. A key feature of the study is related to the modeling of health gains and its inclusion in the objective function, and to be analyzed against the cost dimension. These two objectives are dealt with through the use of the augmented ε-constraint method. Equity considerations in several dimensions (equity of access, equity of utilization, socioeconomic equity and geographical equity) are accounted for in the model through the imposition of satisficing equity levels as model constraints. LTC demand and supply uncertainty are also considered in the model. A case study in Portugal for the 2014-2016 period is used to show the applicability of the model. For the purpose of this case study, health gains are proxied by the Quality Adjusted Life Years (QALYs) gained with the delivery of each type of LTC service. QALYs are a standard measure used to inform resource allocation decisions in the health care sector [16].

The remaining of this paper presents the methodology in Section 2, and the case study and key results in Section 3. Key conclusions and lines for further research are presented in Section 4.

2. Methodology

This study proposes a multi-objective stochastic approach based on a MILP model to support capacity planning and location decisions in the LTC sector. This section starts by presenting background information related to this problem, and then describes the mathematical structure of the model.

2.1. LTC planning background

The location-allocation model proposed in this study supports the (re)organization of a LTC network in the context of a NHS-based country, such as the one that currently operates in Portugal [17]. The model departs from a multi-service network of LTC that ensures the provision of institutional (IC), home-based (HBC) and ambulatory (AC) care services [18]. IC comprise convalescence care (CC), medium-term and rehabilitation care (MTRC), long-term and maintenance care (LTMC) and palliative care (PC), with these different services being characterized by different lengths of stay (LOS). We consider that the state is responsible for establishing contracts with public and private providers so as to ensure the provision of IC, whereas HBC and AC are mainly provided by the state [18].

The planning model provides guidance on i) where and when multiple LTC services should be delivered, ii) how much capacity should be available in each service (both in terms of beds and human resources), iii) how to distribute this capacity across services and patient groups, and iv) which changes are needed in this network over time, including capacity increasing or reduction and the closure and opening of services. These decisions are evaluated in a context of severe budget constraints or limited increase in health care spending, and so the planning of LTC networks needs to consider the minimization of costs. Furthermore, it is a cornerstone of any health system the maximization of health gains for a given level of available resources. The augmented ε-constraint method is employed to explore the trade-off between these conflicting objectives. In this article, other important policy objectives regarding equity – in particular, equity of access (EA), equity of utilization (EU), socioeconomic equity (SE) and geographical equity (GE) – are dealt with as model constraints, assuming that the policy maker accepts achieving equity satisficing levels (following the satisficing concept from Simon [19]).

Since future LTC demand cannot be predicted with total confidence and there is inherent uncertainty in LTC delivery, data uncertainty is modeled either in terms of the number of individuals requiring LTC and in terms of the amount of services required by each of those individuals, as captured by the LOS. A scenario tree approach [20] is selected for handling the uncertainty associated with this uncertain data. Within this two-stage stochastic model, first-stage decisions include the opening and closure of services and investment in new beds, whereas second-stage decisions comprise decisions related to the allocation and reallocation of patients and resources. The objective of the proposed model is thus to minimize expected costs and maximize expected health gains over the uncertain demand scenarios, while simultaneously ensuring the achievement of pre-defined satisficing levels of equity.

2.2. *Mathematical formulation*

2.2.1. Notation. The following indices, sets, parameters and variables will be used.

Indices

d	Demand points	s, p	LTC services	r	Human resources
g	Socioeconomic groups	t, w	Time periods		
l, j	Locations for services	n, k	Scenario tree nodes		

Sets

D	Time periods
$G = G_P \cup G_{NP}$	Socioeconomic groups, divided into subsets G_P (individuals with priority as a result of having lower levels of income) and G_{NP} (other individuals).
$L = L_I \cup L_O$	Locations for services, divided into subsets L_I (locations for IC services) and L_O (locations for HBC and AC services).
R	Human resources
$S = S_I \cup S_O$	LTC services, divided into subsets S_I (IC services) and S_O (HBC and AC services)
T	Time periods
N	Scenario tree nodes
$A = \{(s,r): s\in S, r\in R\}$	Services s provided by human resources r
$F = \{(d,l): d\in D, l\in L\}$	Demand points d that can receive LTC in locations l
$P = \{(k,n): k,n\in N\}$	Predecessors k of the scenario tree node n
$Q = \{(t,n): t\in T, n\in N\}$	Time periods t associated to scenario tree nodes n
$U = \{(s,l): s\in S, l\in L\}$	Services s that can be provided in locations l
$V = \{(s,l): s\in S, l\in L\}$	Services s provided in locations l at t=0
$Z = \{(s,l): s\in S, l\in L\}$	Services s not provided in locations l at t=0

Parameters

D_{dgstn}	Number of individuals from demand point d and socioeconomic group g requiring service s at t in scenario tree node n
gD_{tn}	Number of individuals from lower income groups ($g\in G_P \subseteq G$) requiring LTC at t in scenario tree node n
rD_{dtn}	Number of individuals from demand point d requiring LTC at t in scenario tree node n
uD_{stn}	Number of individuals requiring service s at t in scenario tree node n
eB^0_{sl}	Number of beds available in IC service s ($s\in S_I \subseteq S$) located in l ($l\in L_I \subseteq L$) at t=0
mB_s / MB_s	Minimum/maximum bed capacity allowed per IC service s ($s\in S_I \subseteq S$)
mI_s / MI_s	Minimum/maximum number of individuals allowed per HBC/AC service s ($s\in S_O \subseteq S$)

HRp_{rs}	Hours of care to be provided by human resource r for each individual requiring service s
EA_t / EU_t / GE_t / SE_t	Target defined for EA/EU/GE/SE at t
iC_{st}	Investment cost per new bed installed in IC service s ($s \in S_I \subseteq S$) at t
oC_{st}	Operational cost per service s per period t
rC^1_{st} / rC^2_{st}	Cost of reallocating a bed to IC service s ($s \in S_I \subseteq S$) from a service delivered in a different location/from another service delivered in the same location at t
r_t	Inflation rate at t
LOS_{sn}	Average length of stay (LOS, in days) in IC service s ($s \in S_I \subseteq S$) in scenario tree node n
TP	Number of days per time period
α_{st}	Minimum level of demand that need to be satisfied per service s at t
β	Maximum travel time allowed for patients accessing IC services (in minutes)
θ_{dl}	Travel time between demand point d and service location l (in minutes)
ε_s	Efficiency factor associated with the provision of service s
ρ_n	Probability of scenario tree node n

Variables

X_{slt}	Equal to 1 if service s is located in l at t; 0 otherwise
aB_{slt}	Number of additional beds to invest in for IC service s ($s \in S_I \subseteq S$) located in l ($l \in L_I \subseteq L$) at t
B_{dgsltn}	Number of beds required for individuals from demand point d belonging to socioeconomic group g and receiving IC service s ($s \in S_I \subseteq S$) located in l ($l \in L_I \subseteq L$) at t in scenario tree node n
rB^{in}_{slpjtn} / rB^{out}_{slpjtn}	Number of beds reallocated to/from IC service s ($s \in S_I \subseteq S$) located in l ($l \in L_I \subseteq L$) from/to IC service p ($p \in S_I \subseteq S$) located in j ($j \in L_I \subseteq L$) at t in scenario tree node n
gR_{tn}	Number of individuals from lower income groups receiving LTC at t in scenario tree node n
R_{dgsltn}	Proportion of individuals from demand point d and socioeconomic group g receiving service s in location l at t in scenario tree node n
rR_{dtn}	Number of individuals from demand point d receiving LTC at t in scenario tree node n
uR_{stn}	Number of individuals receiving service s at t in scenario tree node n
$HR_{rdgsltn}$	Number of hours of care provided by human resource r for individuals from demand point d belonging to socioeconomic group g and receiving service s located in l at t in scenario tree node n
U^{tot}_{tn}	Total travel time (in minutes) at t for scenario tree node n
U^{pen}_{tn}	Penalty (in minutes) attributed to individuals not receiving IC at t in scenario tree node n
U^{max}_{tn}	Maximum total travel time (in minutes) at t in scenario tree node n
TIC_{tn} / TOC_{tn}	Total investment/operational cost at t for scenario tree node n

2.2.2. Building the objective function. The two objectives considered include the minimization of expected costs (equations (1-3)) and the maximization of expected health gains (equation (4)). Costs include both investment (equation (2)) and operational costs (equation (3)). Investment costs are related to the investment in new beds (first term in equation (2))) and in the reallocation of beds between services (second and third terms). Operational costs include costs associated with the operation of beds in IC services (first term in equation (3)) and to the provision of HBC and AC services (second term). Health gains vary with the type of service delivered (HG_s) (equation (4)).

$$Min\ f_1 = \sum_{n \in N} \sum_{t:(t,n) \in Q} \rho_n \left(TIC_{tn} + TOC_{tn}\right) \tag{1}$$

$$TIC_{tn} = \sum_{s \in S_I} \sum_{\substack{l \in L_I \\ l:(s,l) \in U}} \left(aB_{slt} \frac{iC_{st}}{(1+r_t)^t} + \sum_{p \in S_I} \sum_{\substack{j \in L_I \\ j:(p,j) \in U \\ j \neq l}} rB^{in}_{slpjtn} \frac{rC^1_{st}}{(1+r_t)^t} + \sum_{\substack{p \in S_I \\ p:(p,l) \in U}} rB^{in}_{slpltn} \frac{rC^2_{st}}{(1+r_t)^t} \right) \quad \forall (t,n) \in Q \tag{2}$$

$$TOC_{tn} = \sum_{d \in D} \sum_{g \in G} \left(\sum_{s \in S_I} \sum_{\substack{l \in L_I \\ l:(s,l) \in U \\ l:(d,l) \in F}} B_{dgsltn} \frac{oC_{st}}{(1+r_t)^t} + \sum_{s \in S_O} \sum_{\substack{l \in L_O \\ l:(s,l) \in U \\ l:(d,l) \in F}} R_{dgsltn} D_{dgstn} \frac{oC_{st}}{(1+r_t)^t} \right) \quad \forall (t,n) \in Q \tag{3}$$

$$Max\, f_2 = \sum_{n \in N} \rho_n \left[\sum_{s \in S} \sum_{t:(t,n) \in Q} HG_s \left(\sum_{d \in D} \sum_{g \in G} \sum_{\substack{l \in L \\ l:(s,l) \in U \\ l:(d,l) \in F}} R_{dgsltn} D_{dgstn} \right) \right] \tag{4}$$

2.2.3. *Defining the constraints of the model.* The proposed model makes use of a set of constraints:

- Opening and closure of services constraints – opening/closing an IC service is not allowed after deciding upon closing/opening it in a previous time period (equations (5-6)). No openings or closures are considered for HBC and AC (equation (7)), given that these services are provided within the scope of a primary care network, which is already established in any NHS-based country;

$$X_{slw} \le X_{slt} \quad \forall s \in S_I, l \in L_I : (s,l) \in (U \cap V), (t,w) \in T, w > t \tag{5}$$
$$X_{slw} \ge X_{slt} \quad \forall s \in S_I, l \in L_I : (s,l) \in (U \cap Z), (t,w) \in T, w > t \tag{6}$$
$$X_{slt} = 1 \quad \forall s \in S_O, l \in L_O : (s,l) \in U, t \in T \tag{7}$$

- Minimum level of demand satisfaction – equation (8) ensures that a minimum level of satisfied demand per service s (captured by the parameter α_{st}) should be guaranteed per time period t.

$$\sum_{n:(t,n) \in Q} \rho_n \left(\frac{uR_{stn}}{uD_{stn}} \right) \ge \alpha_{st} \quad \forall s \in S, t \in T \text{ with } uR_{stn} = \sum_{d \in D} \sum_{g \in G} \sum_{\substack{l:(s,l) \in U \\ l:(d,l) \in F}} R_{dgsltn} D_{dgstn} \quad \forall s \in S, (t,n) \in Q \tag{8}$$

- Single and closest assignment constraints – individuals should receive the care they need in the closest available service (equation (9)), and cannot access services that are not within a maximum travel time (equation (10)). According to equation (9), if location l provides service s and is closer to demand point d than another location j, it means that patients from d should not receive the required care at j;

$$X_{slt} + R_{dgsjtn} \le 1 \quad \forall g \in G, s \in S, (l,j) \in L : (s,l) \in U, (s,j) \in U, (d,l) \in F, (t,n) \in Q, \theta_{dl} < \theta_{dj}, l \neq j \tag{9}$$
$$R_{dgsltn} = 0 \quad \forall g \in G, s \in S, l \in L : (s,l) \in U, (d,l) \in F, (t,n) \in Q, \theta_{dl} > \beta \tag{10}$$

- Resources requirement constraints – number of beds (equation (11)) and human resources (equation (13)) that should be made available per LTC service; and for the particular case of beds, it is also imposed that the bed capacity that needs to be in place in each service and location cannot be less than the bed capacity required in each scenario (equation (12); similarly to [20]). Since in health it is not expected a full occupancy of services, an efficiency factor (ε_s), which takes the value of 1 when occupancy rates are equal to 100%, and lower values for lower occupancy rates, is included in equations (11) and (13);

$$B_{dgsltn} = \frac{LOS_{sn}}{TP} R_{dgsltn} D_{dgstn} \frac{1}{\varepsilon_s} \quad \forall g \in G, s \in S_I, l \in L_I : (s,l) \in U, (d,l) \in F, (t,n) \in Q \tag{11}$$

$$aB_{slt} \geq \begin{cases} \sum_{d\in D}\sum_{g\in G} B_{dgsltn} - eB^0_{sl} - \left[\sum_{p\in S_I}\sum_{\substack{j\in L_I\\ j:(p,j)\in U}}\left(rB^{in}_{slpjtn} - rB^{out}_{slpjtn}\right)\right] & \forall s\in S_I, l\in L_I; (s,l)\in U, (t,n)\in Q, t=1 \\ \sum_{d:(d,l)\in F}\sum_{g\in G}\left(B_{dgsltn} - \sum_{k:(n,k)\in P} B_{dgsl(t-1)k}\right) - \sum_{p\in S_I}\sum_{\substack{j\in L_I\\ j:(p,j)\in U}}\left(rB^{in}_{slpjtn} - rB^{out}_{slpjtn}\right) & \forall s\in S_I, l\in L_I : (s,l)\in U, (t,n)\in Q, t>1 \end{cases} \quad (12)$$

$$HR_{rdgsltn} = R_{dgsltn} D_{dgstn} HRp_{rs}\frac{1}{\varepsilon_s} \quad \forall g\in G, (d,l)\in F, (s,l)\in U, (s,r)\in A, (t,n)\in Q \quad (13)$$

- Minimum and maximum capacity constraints – a minimum and maximum number of beds (equation (14)) and individuals in need (equation (15)) are imposed per service;

$$mB_s X_{slt} \leq \sum_{d:(d,l)\in F}\sum_{g\in G} B_{dgsltn} \leq MB_s X_{slt} \quad \forall s\in S_I, l\in L_I : (s,l)\in U, (t,n)\in Q \quad (14)$$

$$mI_s X_{slt} \leq \sum_{d:(d,l)\in F}\sum_{g\in G} R_{dgsltn} D_{dgstn} \leq MI_s X_{slt} \quad \forall s\in S_O, l\in L_O : (s,l)\in U, (t,n)\in Q \quad (15)$$

- Reallocation constraints – a maximum number of beds is allowed to be reallocated from each IC service (equation (16)); and the number of beds reallocated to service s from service p should be equal to the number of beds removed from service p to service s (equation (17));

$$\sum_{p\in S_I}\sum_{\substack{j\in L_I\\ j:(p,j)\in U}} rB^{out}_{slpjtn} \leq \begin{cases} eB^0_{sl} & \forall s\in S_I, l\in L_I : (s,l)\in U, (t,n)\in Q, t=1 \\ \sum_{d:(d,l)\in F}\sum_{g\in G}\sum_{k:(n,k)\in P} B_{dgsl(t-1)k} & \forall s\in S_I, l\in L_I : (s,l)\in U, (t,n)\in Q, t>1 \end{cases} \quad (16)$$

$$rB^{in}_{slpjtn} = rB^{out}_{pjsltn} \quad \forall (s,p)\in S_I, (l,j)\in L_I : (s,l)\in U, (p,j)\in U, (t,n)\in Q \quad (17)$$

- Equity satisficing levels' constraints – pre-defined satisficing levels of equity are imposed for EA (equation (18)), EU (equation (19)), SE (equation (20)) and GE (equation (21)). The measures selected for defining EA, SE and GE are similar to the ones used in [10], whereas the EU measure is similar to the GE, although ensuring a minimum service level (in comparison to need) across different typologies of LTC services. These satisficing levels should be defined by decision-makers (DMs), corresponding to levels that should be attained (and not optimized) since one is in the context of scarce resources [19].

$$\sum_{n:(t,n)\in Q} \rho_n\left(\frac{U^{tot}_{tn} + U^{pen}_{tn}}{U^{max}_{tn}}\right) \leq EA_t \;\; \forall t\in T \text{ with } \begin{cases} U^{tot}_{tn} = \sum_{d\in D}\sum_{g\in G}\sum_{s\in S_I}\sum_{\substack{l\in L_I\\ l:(s,l)\in U\\ l:(d,l)\in F}} \theta_{dl} R_{dgsltn} D_{dgstn} \;\; \forall (t,n)\in Q \\ U^{pen}_{tn} = \sum_{d\in D}\sum_{g\in G}\sum_{s\in S_I}\beta\left(D_{dgstn} - \sum_{\substack{l\in L_I\\ l:(s,l)\in U\\ l:(d,l)\in F}} R_{dgsltn} D_{dgstn}\right) \;\; \forall (t,n)\in Q \\ U^{max}_{tn} = \sum_{d\in D}\sum_{g\in G}\sum_{s\in S_I}\beta\times D_{dgstn} \;\; \forall (t,n)\in Q \end{cases} \quad (18)$$

$$\sum_{n:(t,n)\in Q} \rho_n\left(1 - \frac{uR_{stn}}{uD_{stn}}\right) \leq EU_t \;\; \forall s\in S, t\in T \quad (19)$$

$$\sum_{n:(t,n)\in Q} \rho_n\left(1 - \frac{gR_{tn}}{gD_{tn}}\right) \leq SE_t \;\; \forall t\in T \text{ with } gR_{tn} = \sum_{d\in D}\sum_{g\in G_P}\sum_{s\in S}\sum_{\substack{l:(s,l)\in U\\ l:(d,l)\in F}} R_{dgsltn} D_{dgstn} \;\; \forall (t,n)\in Q \quad (20)$$

$$\sum_{n:(t,n)\in Q} \rho_n\left(1 - \frac{rR_{dtn}}{rD_{dtn}}\right) \leq GE_t \;\; \forall d\in D, t\in T \text{ with } rR_{dtn} = \sum_{g\in G}\sum_{s\in S}\sum_{\substack{l:(s,l)\in U\\ l:(d,l)\in F}} R_{dgsltn} D_{dgstn} \;\; \forall d\in D, (t,n)\in Q \quad (21)$$

2.3. Solution approach

To deal with the two selected objectives, the augmented ε-constraint method is used [21]. Instead of providing a single optimum solution, this method provides a subset of the Pareto-optimal set. According to Mavrotas [21] the augmented ε-constraint method is a novel version of the conventional ε-constraint method that solves its well-known pitfalls, namely, i) the calculation of the range of each objective function over the efficient set, and ii) the guarantee of efficiency for the obtained solution. For the purpose of this study, the augmented ε-constraint method is applied by minimizing expected costs, whereas the objective related to expected health gains is imposed as constraint. Accordingly, the objective function is now given by equation (22), with μ representing a small number, usually between 10^{-3} and 10^{-6}, and s_2 representing a slack variable; and the health gains-related constraint is given by equation (23), with e_2 depending on the minimum and maximum values for f_2 and on the number of grids points selected for building the Pareto Frontier.

$$Min\ (f_1 - \mu \times s_2) \tag{22}$$

$$f_2 - s_2 = e_2 \tag{23}$$

3. Case-study

Section 3 analyzes the first results from applying the proposed multi-objective stochastic model to the county level in the Great Lisbon region in the 2014-2016 period. It starts by briefly describing how health gains were modeled, and then describes other data in use and presents some results.

3.1. Modeling of health gains

Given that we did not find available data providing estimates of the health gains associated with different LTC interventions, we selected the QALYs gained with the delivery of a LTC service as a proxy for health gains. For estimating QALYs, we used the EQ-5D self-report questionnaire, as it '*has become one of the valuation approaches recommended by several reimbursement authorities and academic bodies in European countries*' [22]. Accordingly, the following steps were used to estimate QALYs:

i. The EQ-5D self-report questionnaire was first completed with information on disabilities for different LTC services [23];
ii. Using the Portuguese EQ-5D value set built by Ferreira et al. [24] together with the information on the EQ-5D level for each type of LTC service, the value of the EQ-5D states before and after receiving LTC within the National Network of Long-Term Care (*Rede Nacional de Cuidados Continuados Integrados*, RNCCI) were calculated;
iii. The QALYs gained per type of LTC service were then determined as the difference between the values of the EQ-5D states after and before receiving LTC.

3.2. Remaining data-set

The model was implemented in the General Algebraic Modeling System (GAMS) 23.7 and was solved with CPLEX 12.0 on a Two Intel Xeon X5680, 3.33GHz computer with 12GB RAM.

The dataset used includes a wide range of information, namely:

- LTC supply at the end of 2013 in the Great Lisbon region [18];
- Operational and investment costs [25];
- Travel time (in minutes) between each county in the Great Lisbon region and each LTC service [26]; and maximum travel time allowed for LTC patients accessing IC services [27];
- Number of hours of care to be provided by physicians and nurses per individual in need [28];
- Efficiency factor associated to the provision of services with a score of 1, given that providers in Portugal are paid according to utilization [18];
- Satisficing equity levels defined by the Head of the Long-Term Care Coordination Team of the Lisbon and Tagus Valley Health Authority, which has given her opinion on these parameters, acting in the role of a real DM in the LTC sector;
- Information about physical, functional, mental and social disabilities of patients receiving care within the RNCCI during 2008, before and after receiving LTC [23].

Journal of Physics: Conference Series **616** (2015) 012007 doi:10.1088/1742-6596/616/1/012007

Regarding uncertainty, the following two parameters were estimated:

i. The number of individuals requiring all the types of IC, HBC and AC, disaggregated by socioeconomic groups (very low income [VLI] and not very low income [NVLI]), as predicted by the detailed simulation model developed in [29];
ii. The LOS associated to each type of IC service estimated with data from [30].

A scenario tree with 81 scenarios was used to describe combinations of these uncertain parameters. To build this scenario tree, the probability distributions associated to both parameters were taken from the outputs of the simulation model developed in [29] and [30], respectively, and were then converted into three annual scenarios using the extended Pearson-Tuckey method [31]. The whole dataset is available from authors, upon request.

3.3. Illustrative results

Figure 1 depicts the Pareto Frontier obtained when running the multi-objective stochastic model to the county level in the Great Lisbon region in the 2014-2016 period. Each solution was obtained by imposing a 120 minutes limit for the computation time, and optimality gaps were below 0.5% for all the solutions. Solution A represents the solution with the minimum expected cost and health gains as measured by QALYs gained, being characterized by the lowest levels of LTC provision. These levels of provision are the minimum levels required to meet the equity satisficing levels indicated by our DM. On the other hand, as we move from solution B to K, the costs of reorganizing and operating the corresponding networks increase, as well as the health gains achieved with LTC provision. The maximum expected health gains and costs are achieved under solution K, with the LTC network found under this solution allowing for full provision of LTC (both IC, HBC and AC).

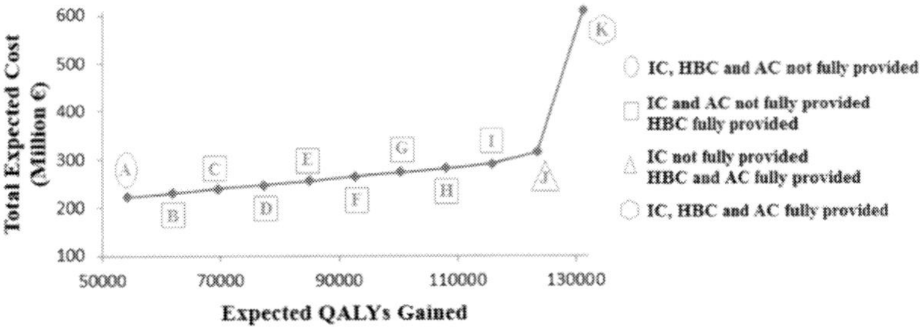

Figure 1 Pareto Frontier obtained when running the proposed multi-objective stochastic model. Legend: IC – Institutional Care; HBC – Home-Based Care; AC – Ambulatory Care.

In order to aid LTC policy makers deciding on how to reorganize the current network of LTC, the cost-effectiveness of each solution compared to the current provision of care is evaluated by determining the additional cost per QALY gained with LTC provision under each solution (also known as Incremental Cost Effectiveness Ratio [ICER] in health technology assessment literature; Table 1).

Table 1. Additional cost (in thousands of euros) per QALY gained, i.e., Incremental Cost Effectiveness Ratios (ICERs), in comparison to current practice for solutions A to K.

Solution	**A**	**B**	**C**	**D**	**E**	**F**	**G**	**H**	**I**	**J**	**K**
ICER	2.83	2.58	2.40	2.25	2.14	2.05	1.97	1.90	1.81	1.86	4.19

According to Table 1, it can be seen that solutions I and J represent the two most cost-effective options (with a lower ICER). These two solutions are characterized by full provision of HBC (see Figure 1) but differ in terms of AC and IC provision. In particular, the LTC network found under solution J ensures full provision of AC, whereas under solution I there is only partial provision of AC. Moreover, comparing these two solutions in terms of IC provision allows for identifying several differences. To illustrate these differences, Figure 2 shows the results obtained for CC services, with CC representing the type of IC service with the highest capacity requirements for both solutions.

Figure 2 provides information on the opening and closure of services, the allocation of individuals to services and the evolution of bed capacity over time for solutions I and J, for the scenario characterized by average demand. Figure 2 shows that, unlike solution I, solution J suggests operating a CC in *Sintra* (1). Also it can be seen that the total CC bed capacity characterizing solution J is higher than in solution I (according to Figure 2, by 2016, there should be 748 and 775 CC beds under solution I and J, respectively). The lower number of services and beds justifies the lower costs and health gains found under solution I when compared to solution J. Note that the higher provision found for CC is related to the high amount of QALYs gained with the provision of these services (the QALYs gained per individual receiving CC, MTRC, LTMC and PC amounts to 0.606, 0.707, 0.315 and 0.214, respectively) and to the higher number of individuals in need for CC, when compared to the remaining IC services.

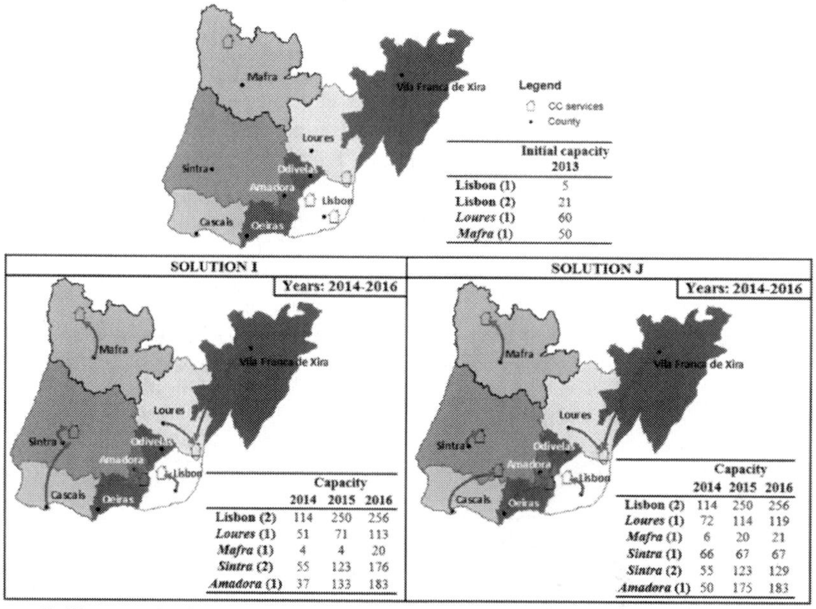

	Initial capacity 2013
Lisbon (1)	5
Lisbon (2)	21
***Loures* (1)**	60
***Mafra* (1)**	50

SOLUTION I

	Capacity 2014	2015	2016
Lisbon (2)	114	250	256
***Loures* (1)**	51	71	113
***Mafra* (1)**	4	4	20
***Sintra* (2)**	55	123	176
***Amadora* (1)**	37	133	183

SOLUTION J

	Capacity 2014	2015	2016
Lisbon (2)	114	250	256
***Loures* (1)**	72	114	119
***Mafra* (1)**	6	20	21
***Sintra* (1)**	66	67	67
***Sintra* (2)**	55	123	129
***Amadora* (1)**	50	175	183

Figure 2 Changes in the provision of convalescence care (CC) services within the LTC network over the 2014-2016 period under solutions I and J (for the scenario characterized by average demand). For simplification purposes, institutions in each county are numbered – e.g. the two institutions in Lisbon are named Lisbon (1) and Lisbon (2).

Table 1 also shows that additional costs per QALY gained decrease as one moves from solution A to I, and increases from solution J onwards. More important than this, comparing the ICERs from A to K with thresholds discussed by some health technology assessment agencies (such as the National Institute for Health and Care Excellence, in the United Kingdom [32]) of around £20000-£30000, it seems that extending the LTC network in the Great Lisbon region in Portugal will be cost-effective.

A final result that can be observed in Figure 1 is that achieving the satisficing equity levels defined by our DM amounts to a minimum of 228 million euros for the three-year period (2014-2016), corresponding to the minimum cost solution (solution A). When compared to the budget available for reorganizing and operating the LTC network in the Great Lisbon region, 30 million euros per year, this value is 250% higher. This result makes it clear that a higher budget is needed for improving the delivery of LTC in the region.

4. Conclusion

One of the top priorities common to the policy agenda of many European countries is related to the improvement in the supply of LTC, and this due to the current ageing phenomenon, as well as to the increasing prevalence of chronic diseases. Furthermore, to deal with the strong pressures that are currently in place to decrease public health care spending, the planning of networks of care requires accounting for cost, equity and health gains related objectives.

Within this context, a multi-objective two-stage stochastic MILP model is proposed to support location-allocation decisions in the LTC sector under the context of NHS-based countries. The model aims at informing the (re)organization of multi-service networks of LTC while maximizing expected health gains and minimizing expected costs (with satisficing levels of equity in several dimensions being achieved), with health gains representing a dimension usually ignored in location-allocation studies. The trade-off between these conflicting objectives is addressed through the use of the augmented ε-constraint method. A case study in the Great Lisbon region in Portugal for the 2014-2016 period is explored. The estimation of health gains was done through the computation of QALYs.

Key results from the model application confirm that the budget currently available for improving LTC delivery in the Great Lisbon region is far from being able to achieve the satisficing equity levels defined by the DM; but investments in the LTC network are cost-effective. Results allowed for exploring the trade-off between cost and health gains, showing the relevance of considering both objectives when planning a network of LTC; and information on the incremental cost effectiveness ratio is particularly useful for policy makers in the LTC sector deciding on how to plan networks of care.

Several research topics should be pursued as further work. First, if a single solution is sought, rather than multiple solutions, after generating the Pareto optimal solutions, one may resort to Multiple Criteria Decision Making methods and use the generated solutions as discrete alternatives so as to assist decision makers selecting the most preferred solution. It is also relevant to explore the impact of adopting different health policy options in planning decisions. For instance, allowing for substituting institutional care by home-based care may result in higher health gains for the same level of cost, due to the lower QALYs gained with the provision of certain types of institutional services. Furthermore, alternative ways to capture benefits in the delivery of care can be explored. For instance, measures of social welfare – which are more holistic than health gains – can be explored.

Acknowledgments

The first author acknowledges financing from the *Fundação para a Ciência e Tecnologia* (Portugal) (SFRH/BPD/98270/2013). The authors thank the support from the Institute of Operations Research, Karlsruhe Institute of Technology (KIT, Germany) and from the Centre for Management Studies of *Instituto Superior Técnico* (CEG-IST, Portugal). The authors thank Dr. Regina Sequeira Carlos, the Head of the Long-Term Care Coordination Team of the Lisbon and Tagus Valley Health Authority, for her willingness to take part in this study.

References

[1] Brodsky J and Clarfield AM 2009 Organization of health services: long term care in health services *Health systems policy, finance, and organization* ed G Carrin, K Buse, K Heggenhougen and SR Quah (San Diego: Academic Press) pp 351-357.

[2] World Health Organization 2000 *Home-based long-term care: report of a WHO study group* (Geneva: World Health Organization).

[3] Brailsford S and Vissers J 2011 OR in healthcare: A European perspective *Eur J Oper Res.* **212** 223–234.

[4] Rahman S-u and Smith DK 2000 Use of location–allocation models in health service development planning in developing nations *Eur J Oper Res.* **123** 437-452.

[5] Smith HK, Harper PR and Potts CN 2012 Bicriteria efficiency/equity hierarchical location models for public service application *J Oper Res Soc.* **64** 500-512.

[6] Abdelaziz FB and Masmoudi M 2012 A multiobjective stochastic program for hospital bed planning *J Oper Res Soc.* **63** 530-538.
[7] Mestre AM, Oliveira MD and Barbosa-Póvoa AP 2015 Location–allocation approaches for hospital network planning under uncertainty *Eur J Oper Res.* **240** 791–806.
[8] Powell M and Exworthy M 2003 Equal access to health care and the British National Health Service *Policy Stud.* **24** 51-64.
[9] McKenna C, Chalabi Z, Epstein D and Claxton K 2010 Budgetary policies and available actions: A generalisation of decision rules for allocation and research decisions *J Health Econ.* **29** 170-181.
[10] Cardoso T 2013 Methods to Inform the Management and Planning of Long-Term Care Services (Department of Engineering and Management, Instituto Superior Técnico, Universidade de Lisboa, Lisbon).
[11] Chalabi Z, Epstein D, McKenna C and Claxton K 2008 Uncertainty and value of information when allocating resources within and between healthcare programmes *Eur J Oper Res.* **191** 530–539.
[12] Owen SH and Daskin MS 1998 Strategic facility location: A review *Eur J Oper Res.* **111** 423-447.
[13] Cardoso T, Oliveira MD, Barbosa-Póvoa A and Nickel S 2013 Planning a long-term care network with uncertainty, strategic policy and equity considerations: A stochastic planning approach (Working Paper CEG-IST no. 4/2013, Instituto Superior Técnico).
[14] Santibáñez P, Bekiou G and Yip K 2009 Fraser Health uses mathematical programming to plan its inpatient hospital network *Interfaces.* **39** 196-208.
[15] Lin F, Kong N and Lawley M 2012 Capacity planning for publicly funded community based long-term care services *Community-based operations research* vol 167 ed MP Johnson (New York: Springer) pp 297-315.
[16] Drummond MF, Sculpher M, Torrance G, O'Brien B and Stoddart G 2005 *Methods for the economic evaluation of health care programmes* (Oxford: Oxford University Press).
[17] Barros PP, Machado SR and Simões JA 2011 Portugal: Health system review *Health systems in transition.* **13** 1-156.
[18] Mission Unit for Integrated Continuing Care 2013 Unidade de Missão para os Cuidados Continuados Integrados [Mission Unit for Integrated Continuing Care] URL: www.rncci.min-saude.pt.
[19] Simon HA 1956 Rational Choice and the Structure of the Environment *Psychol Rev.* **63** 129–138.
[20] Birge JR and Louveaux F 1997 *Introduction to stochastic programming* (New York: Springer Verlag).
[21] Mavrotas G 2009 Effective implementation of the e-constraint method in Multi-Objective Mathematical Programming problems *Appl Math & Comput.* **213** 455–465.
[22] Szende A, Oppe M and Devlin N 2007 *EQ-5D value sets: inventory, comparative review and user guide* (The Netherlands: Springer).
[23] Rato H, Rodrigues M and Rando B 2009 Estudo de caracterização dos utentes da Rede Nacional de Cuidados Continuados Integrados – Relatório Final [Study characterizing the patients from the National Network of Long-Term Care - Final Report] (Oeiras: Instituto Nacional de Administração).
[24] Ferreira LN, Ferreira PL, Pereira LN and Oppe M 2014 The valuation of the EQ-5D in Portugal *Qual Life Res.* **23** 413-423.
[25] Portuguese Observatory on Health Systems 2011 Relatório de Primavera 2011 [2011 Spring Report] (OPSS).
[26] ViaMichelin 2013 ViaMichelin Maps and Route Planner 2013, http://www.viamichelin.co.uk/.

[27] Health Regulation Authority 2011 Estudo do acesso dos utentes aos cuidados continuados de saúde [Study on patients' access to long-term care] (Oporto).

[28] ERS 2013 Avaliação do acesso dos utentes aos cuidados continuados de saúde [Evaluation of the access of patients to long-term care] (Entidade Reguladora de Saúde).

[29] Cardoso T, Oliveira M, Barbosa-Póvoa A and Nickel S 2012 Modeling the demand for long-term care services under uncertain information *Health Care Manag Sci.* **15** 385-412.

[30] Central Administration of the Health System 2013 Implementação e Monitorização da Rede Nacional de Cuidados Continuados Integrados - Relatório Final [Implementing and Monitoring the National Network of Long-Term Care - Final Report] (Cuidados Continuados - Saúde e Apoio Social).

[31] Clemen RT and Reilly T 2003 Making Hard Decisions with Decision Tools Suite Update 2004 (Duxbury).

[32] NICE 2014 National Institute for Health and Care Excellence - The Guidelines Manual http://www.nice.org.uk/article/pmg6/chapter/7-assessing-cost-effectiveness.

Journal of Physics: Conference Series **616** (2015) 012008 doi:10.1088/1742-6596/616/1/012008

Stochastic optimization of the scheduling of a radiotherapy center

Antoine Legrain[1], Marie-Andrée Fortin[2], Nadia Lahrichi[1], Louis-Martin Rousseau[1], and Marino Widmer[3]

[1]CIRRELT and Polytechnique Montréal, Department of Mathematical and Industrial Engineering, C.P. 6079, Succursale Centre-ville, Montréal, QC, Canada, H3C 3A7,

[2]Head of the Radio-Oncology Department, CICL and Department of Radiology, Radio-Oncology and Nuclear Medicine, Université de Montréal, C.P. 6128, Succursale Centre-ville, Montréal, QC, Canada, H3C 3J7, and

[3]DIUF Decision Support & Operations Research, University of Fribourg, Fribourg, Switzerland

E-mail:
`antoine.legrain@polymtl.ca, mfortin.csssl@ssss.gouv.qc.ca, nadia.lahrichi@polymtl.ca, louis-martin.rousseau@polymtl.ca, and marino.widmer@unifr.ch`

Abstract. Cancer treatment facilities can improve their efficiency for radiation therapy by optimizing the utilization of the linear accelerators (linacs). We propose a method to schedule patients on such machines taking into account their priority for treatment, the maximum waiting time before the first treatment, the treatment duration, and the preparation of this treatment (dosimetry). At each arrival of a patient, the future workloads of the linacs and the dosimetry are inferred. We propose a genetic algorithm, which schedules future tasks in dosimetry and a constraint programming formulation to verify the feasibility of a planning of dosimetry. This approach ensures the beginning of the treatment on time and thus avoids the cancellation of treatment sessions on linacs. Preliminary results indicate the improvements of this new procedure.

1. Introduction

Patients with cancer can wait significant amount of time in Quebec for their treatments. After the first consultation, a patient must take a series of exams before undergoing his treatment on a linear accelerator (linac), which irradiates the malignant tumor to kill the infected cells. The total waiting delay is measured as the number of waiting days between the first consultation and the beginning of the treatment. Managers need to reduce this delay and must report detailed statistics to state authorities on a regular basis. For example, a team of the McGill University Health Center in Montreal has decreased the delay from 76 days to 56 days in one year for patients suffering from cell lung cancers [1]. The team achieves such a goal by improving their processes and their patient flow. However, Quebec authorities have fixed an objective of 28 days as a maximum delay. There still need improvements for cancer care processes: this improvements are crucial knowing that cancer costs $17.4 billion per year in Canada and that the number of patients diagnosed with cancer keeps increasing by 1.5% every year [2]. Legrain *et al* [3] have shown that an efficient utilization of the linacs can decrease this delay. Despite this improvement, the patient can face some unnecessary delays if the preparation of the cancer treatment (pretreatment) is not correctly planned and executed. The pretreatment consists mainly in planning the shape, the intensity, and the direction of the beams of the linac: this part is called the dosimetry. If pretreatment planning is not completed, the beginning of the treatment can be postponed or even canceled. Indeed, the appointments on the linac are usually taken before the end of the pretreatment, and thus a delay can cancel the first ones. This paper presents tools to improve the efficiency of the pretreatment process. It aims at reducing pretreatment process times and at avoiding unnecessary cancellations.

This work has been realized in collaboration with the Centre Intégré de Cancérologie de Laval (CICL) and is based on previous work [3]. The main bottlenecks are first the linacs and second the dosimetry. We propose to deal both bottlenecks in an online fashion. When a patient arrives in the center, a clerk books his appointments on a linac according to the number of prescribed sessions. The challenge lies in evaluating the future resource needs to book the best possible appointments. A cancer treatment facility must treat two different kind of patients: palliatives and curatives. Palliative patients need a quick treatment to relieve their pain, they are considered as high priority patients and should start their treatment less than 3 days after their admission. The center delivers a more precise treatment to curative patients to maximize their recovery chances; they are considered to be lower priority patients and can wait until 14 or 28 days according to their type of cancer. The management should balance their resources between these two kinds of patients and must respect as much as possible these deadlines. The pretreatment is simplified and reduced to a series of fixed waiting times and two tasks in dosimetry. The first task is the preparation of the treatment and the second task is its verification. The two tasks must be performed by two different dosimetrists. The fixed delays allow to perform all the others tasks of the pretreatment. The scheduling of the pretreatment can thus be viewed as an hybrid flow-shop with recirculation.

Job-shop scheduling problems are known to be difficult to solve. Methods such as mixed-integer programming [4] or constraints programming [5] achieve poor performance on big and complex job-shops. Gélinas and Soumis [6] present a column generation approach which succeeds to solve problems with 250 jobs only if there is less than two operations by jobs. When the number of operations increases and the constraints become more complex, one prefers to use heuristics or meta-heuristics. Widmer and Hertz [7] propose a successful tabu search for a flow-shop problem; Cheng *et al* [8] give an important review of genetic algorithms applied to job-shop scheduling. Gröflin *et al* [9] also propose advanced tabu search to solve very complex job-shop scheduling problems. The flow-shop problem is a little easier, because operations in a job are always performed in the same order. Furthermore, when an operation can be realized on different or parallel machines, the problem is so-called the hybrid flow-shop problem. Ruiz and Vázquez-Rodríguez [10] present an extensive review of this problem. As the dosimetry process imposes that the preparation and the verification operations are performed by the same team of dosimetrists, this problem is modeled by an hybrid flow-shop with recirculation. Bertel and Billaut [11] provide a very efficient genetic algorithm with an interesting cyclic crossover mechanism.The genetic algorithm introduced here draws inspiration from this latter procedure.

Pretreatment scheduling does not have an extensive literature. Castro and Petrovic [12] present a mathematical programming formulation and some heuristics to solve the whole pretreatment scheduling problem. They assume that the appointments on linacs have already been booked: the pretreatment planning must thus be built considering these appointments as hard due dates. The contribution of this paper is thus to propose a matheuristic solving the pretreatment scheduling and the radiotherapy booking problems at the same time in an online fashion: it is a flow-shop problem with two kind of machines.

Section 2 presents the whole formulation and algorithms. Section 3 shows preliminary results on a real instance from the CICL. Section 4 closes the article with final remarks.

2. Methodology

The model proposed by Legrain *et al* [3] is here modified to take into account the dosimetry scheduling. The formulation (1) presents a stochastic optimization model. The problem is modeled as a two-machines flow shop: the dosimetry and the linacs. The planning of the dosimetry and appointments on linacs are represented by columns. At the arrival of patient j, the model infers the average cost of linacs' planning on a set Ω_j of future patients scenarios. The variables x_{ij}, y_{il}^{ω}, and v_i^{ω} are equal to 1 if respectively the linac's appointments $i \in S_j$ is chosen for patient j, the linac's appointments $i \in S_l$ is chosen for patient l in the scenario $\omega \in \Omega_j$, and the dosimetry planning $i \in S^D$ is chosen for all patients of the scenario $\omega \in \Omega_j$. These variables are equal to 0 otherwise. The linacs appointments are represented by the parameters a_{ilk}^{m} $(= 1$ if the patient l is booked on linac m the day k for the appointments i, 0 otherwise), b_{ij} (the day of the first appointment for the appointments i) and the cost c_{ij}. The price c_{ij} of a linac appointments is a combination of waiting times and deadlines violation costs. The end of the pretreatment for patient l in the dosimetry planning i is the parameter r_{ij}. Finally, the variables z_{mk}^{ω} count the number of slots booked in overtime the day k on linac m. They are bound for the linac m by O_{day} on a day and by O_{week} on a week.

$$\min \sum_{i \in S_j} c_{ij} x_{ij} + \mathbb{E}_{\omega \in \Omega_j} [\sum_{l \in \mathcal{P}^{\omega}} \sum_{i \in S_l} c_{il} y_{il}^{\omega} + \sum_{k \in H} \sum_{m \in M} c^{o} z_{mk}^{\omega}] \tag{1a}$$

subject to:

$$\sum_{i \in S_j} x_{ij} + \sum_{i \in S_l} y_{ij}^{\omega} \geq 1 \tag{1b}$$

$$\sum_{i \in S_l} y_{il}^{\omega} \geq 1, \qquad \forall \omega \in \Omega_j, \forall l \in \mathcal{P}^{\omega} \tag{1c}$$

$$\sum_{i \in S^D} v_i^{\omega} \geq 1, \qquad \forall \omega \in \Omega_j \tag{1d}$$

$$\sum_{i \in S_j} b_{ij} x_{ij} + \sum_{i \in S_l} b_{ij} y_{ij}^{\omega} - \sum_{i \in S^D} r_{ij} v_i^{\omega} \geq 0, \qquad \forall \omega \in \Omega_j \tag{1e}$$

$$\sum_{i \in S_l} b_{il} y_{il}^{\omega} - \sum_{i \in S^D} r_{il} v_i^{\omega} \geq 0, \qquad \forall \omega \in \Omega_j, \forall l \in \mathcal{P}^{\omega} \tag{1f}$$

$$\sum_{i \in S_j} a_{ijk}^{m} x_{ij} + \sum_{l \in \mathcal{P}^{\omega}} \sum_{i \in S_l} a_{ilk}^{m} y_{il}^{\omega} \leq F_k^m + z_{mk}^{\omega}, \qquad \forall m \in M, \forall k \in H, \forall \omega \in \Omega_j \tag{1g}$$

$$\mathbb{1}_{\mathcal{P}_p}(j) \sum_{i \in S_j} a_{ijk}^{m} x_{ij} + \sum_{l \in \mathcal{P}_p^{\omega}} \sum_{i \in S_l} a_{ilk}^{m} y_{il}^{\omega} \geq z_{mk}^{\omega}, \qquad \forall m \in M, \forall k \in H, \forall \omega \in \Omega_j \tag{1h}$$

$$\sum_{k=b}^{b+4} z_{mk}^{\omega} \leq O_{week}, \qquad \forall m \in M, \forall b \in \mathcal{B}, \forall \omega \in \Omega_j \tag{1i}$$

$$z_{mk}^{\omega} \in [0, O_{day}], \qquad \forall m \in M, \forall k \in H, \forall \omega \in \Omega_j \tag{1j}$$

$$x_{ij} \in \{0,1\}, \qquad \forall i \in S_j \tag{1k}$$

$$y_{il}^{\omega} \in \{0,1\}, \qquad \forall l \in \mathcal{P}^{\omega}, \forall i \in S_l, \forall \omega \in \Omega_j \tag{1l}$$

$$v_i^{\omega} \in \{0,1\}, \qquad \forall i \in S^D, \forall \omega \in \Omega_j \tag{1m}$$

Constraints (1b) and (1c) ensure respectively that patient j and all future patients are scheduled. Constraints (1d) choose a dosimetry schedule for each scenario ω. Constraints (1e) and (1f) ensure respectively that patient j and all future patients have their pretreatment ready on time for their first treatment on a linac: they are precedence constraints from the flow-shop problem. Constraints (1g) verify that the capacity (including the overtime) of each linac is not exceeded. Constraints (1h) force palliative patients to be scheduled in available overtime slots. Constraints (1i) and (1j) bound respectively the weekly and daily overtime on each linac. Constraints (1k), (1l), and (1m) are integrality constraints. Finally, the objective (1a) is divided in two parts: the cost of a planning for the patient j and the average future cost of linacs' planning.

The columns x_{ij}, y_{il}^{ω} representing patients appointments on linacs are all inserted once at the beginning. The others columns v_i^{ω} are generated during the process with the genetic algorithm 1 inspired by Bertel and Billaut [11].

Algorithm 1 Genetic Algorithm

Population: list of chromosomes: $P := [\,]$
Initialization: add $N/4$ chromosomes built with dispatching rules, fill the rest with random chromosomes
while solving time $< T$ AND not enough different chromosomes with a negative reduced cost **do**
  for all chromosome c in population P **do**
    Cyclic Crossover: with probability p_c, cross c with a random different chromosome
    Mutation: with probability p_m, swap randomly two positions of c
    Intensification: with probability p_l, make a small local search with insertions on c
  end for
  Add all new chromosomes to P
  Selection: keep the best N chromosomes
end while

The dosimetry scheduling problem has two tasks and a time window per job. The time windows correspond to some fixed delays: they ensure that all the others pretreatment tasks than dosimetry's ones are performed within this waiting times. The two tasks represent the preparation and the verification of the treatment by a dosimetrist. As the formulation (1) follows a Dantzig-Wolfe scheme [6], this flow-shop problem minimizes the weighted completion time (due to the dual variables linked to constraints (1e) and (1f)). Algorithm 1 tries thus to build in a certain amount of time T a population P of N chromosomes with a negative reduced cost. A chromosome is a sequence of jobs and represents a dosimetry planning. A job can be found two times in a chromosome: the first appearance corresponds to the preparation of the dosimetry, while the second corresponds to the verification. A planning is then made from a chromosome by simply scheduling in the chromosome order each task as early as possible. Finally, the intensification phase aims at improving some chromosomes: several insertions are tested to decrease the reduced cost.

The algorithm 2 presents the global booking procedure for each patient j. We solve the slave problems obtained by the Benders' decomposition [13] applied on (1) instead of solving the whole formulation. A dispatching rule schedules all the pending dosimetry tasks to build an operational planning. Algorithm 2 uses specifically the earliest late start time rule which gives a good solution for a parallel machine problem [14]. Yet the feasibility of a set of pretreatment tasks is checked with a constraints programming model. Finally, the computation of the average reduced costs is explained in [3].

Journal of Physics: Conference Series **616** (2015) 012008 doi:10.1088/1742-6596/616/1/012008

Algorithm 2 Online Scheduling Algorithm

Initialize waiting list of dosimetry operations: $Q := [\,]$
for all patient j arrivals **do**
  schedule dosimetry tasks until current patient arrival day with a dispatching rule
  remove all scheduled dosimetry's operations from Q
  for all scenarios ω **do**
    solve slave problems with value $x = 0$ for event ω
    while genetic algorithm adds new columns **do**
      insert new columns
      solve slave problems for event ω
    end while
  end for
  select a schedule i which is:
  • **minimum** for the average reduced cost
  • **feasible** for all dosimetry's operations in Q (checked with constraints programming)
  update remaining resources F_k^m
  insert the two dosimetry tasks of patient j in Q
end for

3. Results

The instance used for the computational test is a real data set of the CICL with 181 patients on 77 working days. There are two linacs with 29 slots plus 3 in overtime and 4 dosimetrists. The scenarios used for the online algorithms have been drawn from the empirical patients distribution of the CICL: 5 scenarios are sampled for each patient. However, a proportion (70%) of curative patients are known in advance as they have undergone a surgery and/or a chemotherapy treatment in the center.
Table 1 compares four algorithms according to different criteria: the number of violated deadlines, the average waiting time, the number of used overtime slots, the number of patients which have respectively violated a deadline or seen their first appointment canceled because of the pretreatment. The CICL algorithm is the one currently used by the facility. The online clairvoyant algorithm is presented in [3]. These two procedures, unlike the two following, do not take into account the dosimetry planning in their decision process, but just a fixed delay. The online scheduling clairvoyant algorithm follows the procedure 2. Finally, the offline procedure proposes the optimal solution of the problem by solving it once at the end when all patients are known.

Table 1. Comparison of different algorithms on a real instance.

Algorithm	Deadline >3 days	Deadline >14 days	Deadline >28 days	Average delay 3 days	Average delay 14 days	Average delay 28 days	Overtime slots	Appointment delayed	Appointment canceled
CICL	15	4	0	1.54	9.20	8.63	17	0	2
Online Clairvoyant	5	0	0	0.80	8.02	10.37	3	1	4
Online Scheduling Clairvoyant	1	0	0	0.26	7.70	11.44	0	0	0
Offline	1	0	0	0.61	7.28	11.57	0	0	0

Table 1 shows that the offline and online scheduling clairvoyant algorithms outperform the two others procedures. They are nearly perfect with just 1 deadline violation, 0 overtime slots and 0 appointment delayed or canceled. The online clairvoyant algorithm has just 5 deadline violations and 3 overtime slots. However, one patient has been delayed because of the pretreatment and 4 patients pretreatments have not been ready on time to start the treatments. This last point means that the fixed delay was not enough to finish the pretreatment on schedule. The CICL algorithm cannot achieve to respect the deadlines as 19 deadlines are violated. Finally, the procedure 2 needs in average 15 seconds per patient to book his appointments.

4. Conclusion

We propose to model and solve a complex radiotherapy stochastic scheduling problem in an online fashion. A stochastic and complex column generation model is used for each patient. The relaxations made in our previous work decrease the solving time. The genetic algorithm allows also to reach this reasonable solving time by generating good columns even if the underlying flow-shop problem is extremely difficult. These preliminary results show that our online scheduling clairvoyant algorithm achieves results comparable to the optimal solution. Our procedure also outperforms the others proposed online algorithms. For our future works, we aim at providing a deep sensitivity analysis and more extensive computational tests.

References

[1] Cameron D 2012 Cancer: le dr mulder veut réduire les délais de traitement URL `http://www.lapresse.ca/actualites/sante/201207/17/01-4556525-cancer-le-dr-mulder-veut-reduire-les-delais-de-traitement.php`

[2] Canadian Cancer Society's Advisory Committee on Cancer Statistics 2014 Canadian cancer statistics 2014 Tech. rep. Canadian Cancer Society Toronto, ON

[3] Legrain A, Fortin M A, Lahrichi N and Rousseau L M 2014 Online stochastic optimization of radiotherapy patient scheduling *Health care management science* **3** 1–14

[4] Desaulniers G, Desrosiers J and Solomon M M 2005 *Column generation* vol 5 (Springer)

[5] Baptiste P, Le Pape C and Nuijten W 2001 *Constraint-based scheduling: Applying constraint programming to scheduling problems* vol 39 (Springer)

[6] Gélinas S and Soumis F 2005 *Dantzig-Wolfe decomposition for job shop scheduling* (Springer)

[7] Widmer M and Hertz A 1989 A new heuristic method for the flow shop sequencing problem *European Journal of Operational Research* **41** 186–193

[8] Cheng R, Gen M and Tsujimura Y 1996 A tutorial survey of job-shop scheduling problems using genetic algorithms–I. representation *Computers & Industrial Engineering* **30** 983–997

[9] Gröflin H, Pham D N and Bürgy R 2011 The flexible blocking job shop with transfer and set-up times *Journal of combinatorial optimization* **22** 121–144

[10] Ruiz R and Vázquez-Rodríguez J A 2010 The hybrid flow shop scheduling problem *European Journal of Operational Research* **205** 1–18

[11] Bertel S and Billaut J C 2004 A genetic algorithm for an industrial multiprocessor flow shop scheduling problem with recirculation *European Journal of Operational Research* **159** 651–662

[12] Castro E and Petrovic S 2012 Combined mathematical programming and heuristics for a radiotherapy pre-treatment scheduling problem *Journal of Scheduling* **15** 333–346

[13] Benders J F 1962 Partitioning procedures for solving mixed-variables programming problems *Numerische mathematik* **4** 238–252

[14] Davis E W and Patterson J H 1975 A comparison of heuristic and optimum solutions in resource-constrained project scheduling *Management science* **21** 944–955

Operations Research on Hospital Admission Systems: a first overview of the 2005-2014 decade

A Teixeira[1,2] and M J F De Oliveira[3]
[1] Department of Mathematics, University of Trás-os-Montes e Alto Douro, 5000 Vila Real, Portugal.
E-mail: ateixeir@utad.pt
[2] CIO–Operations Research Center, FCUL - University of Lisbon, 1749 Lisbon, Portugal
[3] Federal University of Rio de Janeiro / Centro de Tecnologia, Bloco F sala 105, Caixa Postal 6851, Rio de Janeiro, RJ Brazil.
E-mail: mario_jo@po.coppe.ufrj.br

Abstract. The main characteristics of researches on Operations Research (OR) applied to Hospital Admission Systems (HAS) are described and analysed, not only in order to determine the interest of this topic and its evolution over the last decade, but also to improve our understanding on this issue. A database with information on this theme was built, considering relevant publications in OR applied to HAS, within the 2005–2014 period. The information contained in those works was catalogued in accordance with specific key elements previously chosen. Additionally, the developments on application of OR technics to HAS in the last decade are evaluated, observing what has been done, describing the related applications and procedures, and their practical implementation. The result is a review, providing a new insight on the recent enhancements of OR technics applied to HAS and trying to outline other pathways for the near future.

1. Introduction

The first applications of OR to hospitals dates from the late 1940´s and are associated with the development of the National Health Services in Great Britain [42]. In the United States of America, the homologous situation happened as a consequence of Hill-Burton Act in 1946, given that the creation of a planning program that would survey existing facilities and promote hospital construction plans was required [58]. Since then, many specific topics in this theme appeared, namely "Regional Planning", "Estimation of Regional Needs" and "Hospital Activities", "Statistical Patterns of demand for hospital services", "Hospital facilities utilization", "Intensity of Medical Care", "Length of Stay in hospitals" and "Economics of automating selected hospital activities" [1], among others.

The concern with demand and resource utilization in hospitals sparked the interest in Hospital Admission Systems (HAS) or at least in their components, which has become a subject of research interest since the 1950s, both in UK and in USA. Such problems have attracted the attention of operational researchers such as Milsum [40], Bailey [5], Balintfy [6], Thompson [54], Connor [19], Blumberg [10] and Young [57].

Although it is possible to find, in the literature, valuable and interesting synthesis and reviews on the application of Operations Research (OR) technics to health care systems or to their components, namely the ones by Boldy [8], Brailsford and Vissers [11], Campbell [14], Cardoen et al. [15], De Oliveira [20], Fries [24], Günal and Pidd [28], Jack [33], Lagergren [36] and Royston [44], the authors of the present paper consider that further analysis on this subject is still needed, not only in order to provide better guidance to hospital staff, but also to provide additional information to policy makers and to all the entities involved in the health care system; moreover, due to the economic crisis framework in which the world stands now.

[1] To whom any correspondence should be addressed.

The purpose of this paper is to present some of the work that has been developed concerning the theme of OR applied to HAS. An overview on the works on this theme is outlined, in order to better understand what has been done in this area, to gain insight on the future directions of this field and, thus, draw some improvements from the most recent developments. Here, we center our attention on the works developed since 2005.

The structure of the paper considers: in Section 2, the methodology of the study is presented; in Section 3, the results of this overview are described; and in Section 4, the main conclusions are presented.

2. Methodology of the study

Although some works about enhancements on OR technics applied to Health Systems have been published during the last decade, the authors opinion is that further developments and applications are still in need. The aim of this study is to analyze the main features on the concept of HAS from a series of relevant manuscripts that have been disseminated in some of the most prestigious publications in the last decade; characterize and aggregate the information contained in these works according to pre-specified issues and, finally, gather it in a database.

This literature analysis is quite large and it must be delimited, so that the quantity and diversity of data do not prevent us from achieving our purposes. Health care is an enormous field and there are many different ways of analyzing it. We should point out that we are only interested in those works that can be considered essential references in the HAS field. Therefore, a set of relevant databases and web search engines is selected, namely Web of Science, Current Contents Connect, Inspec, PubMed, Academic Research Microsoft and Scholar Google, since they contain a wide range of top journals that report research findings. Here, the focus goes essentially to specialized works in the OR and Health field published on books, journals, proceedings, as well as MSc and Ph.D. thesis and working papers, completing the selection with other relevant publications in this area. This selection was made in the understanding that published and unpublished quality works tend to be presented in this type of manuscripts, which is needed so that our findings will be based on research and innovation developments that are observing good value criteria. It is worth noting that similar process and sources of information were used by Cardoen et al. [15], Gonseth et al. [27] and Tam et al. [51], among others. Additionally, in order to increase the reading, most of the considered manuscripts in the database are written in English, being only a very small portion of them written in French and Portuguese.

We have been building a database of publications in OR applied to HAS. The information in these works was cataloged in accordance with specific key elements previously chosen taking into account the topics that have been explored in the literature over the past six decades and also the methodological approach suggested by several authors. All the works are described in terms of the following eleven key elements, which also represent the components involved in a real system:

- Entities involved, namely: administrators, auxiliary personnel, doctors, nurses, patients, psychologists and technicians.
- Activities performed, such as: admit, allocate, control, discharge, evaluate, inform, manage, monitor, plan, predict, schedule and transfer.
- Resources used, like: beds, equipment, financial, rooms, surgery room, units and wards.
- Services involved, such as: ambulatory, blood bank, emergency, intensive care, pharmacy, radiology, surgery, tomography and X-ray.
- Level of planning: geographic (local, national, regional), political, strategic and tactical.
- Type of Hospital: acute, emergency, general, outpatient, private, public and teaching.
- Type of Patient: booked, elective, emergency, inpatient and outpatient.
- Performance measures, like: accessibility, capacity, demand, efficiency/performance, length of stay, quality of care, satisfaction, utilization rate, waiting time and work load.
- Type of model, for example: analytic, computational, continuous, decision, deterministic, discrete and stochastic.

- OR methodology applied, for instance: Heuristics, Mathematical Programming, Optimization, Queuing Theory, Simulation and Stochastic Process.
- Theory / practice: Comments on theoretical and/or practical implementation of the results.

Moreover, the developments on the application of OR technics to HAS in the last decade are evaluated, observing what has been done, describing the related applications and procedures, and exploring their practical implementation.

We carried out a manuscript search using the terms "Hospital", "Admission" and "Systems", which we feel that would be the three keywords most likely to suit our purpose. Initially, we began by performing a search using the three previous terms in the title, abstract or keywords of the analyzed manuscripts. A manual search was also performed to detect synonyms of the above expressions. In the next phase, the references cited in the manuscripts obtained, as well as other items listed in the search engines, were also examined and, when appropriate, included in the database. After this process, the authors examined individually the information contained in each article and the most important elements were identified. If there was a discrepancy in the categorization of articles reviewed a reassessment was performed until there was a consensus.

3. Results of the study

Regarding the last decade, about 261 new manuscripts on HAS were found in the literature, which demonstrates the interest of the researchers in this theme. Table 1 contains a summary of the inclusion/exclusion procedure, namely the total number of articles found in each one of the databases / web search engines used, the number of duplicated works and the number of articles left, after excluding the ones which did not use any OR technics. Among all the works found, our focus goes to the 90 ones which use OR technics, wherein more than 75% of them are articles published in scientific journals, slightly more than 14% are conference proceedings and between 2% and 4% correspond to working papers, Phd Thesis and Books. To reveal a wealth of new material in a succinct way would be a very difficult task, while trying to do justice to all the individual authors in such short number of pages would be impossible. Therefore, we decided to present here a synthesis on the recent progress and only describe some of the contributions of the analyzed works, mainly in terms of the most used OR technics, aiming to present a more detailed review on this topic in a near future.

Table 1. Summary of the study selection

Sources	Articles found	Duplicated	Non rejected
Academic Research Microsoft	16	3	6
Inspec	1	0	0
Pub Med	17	3	2
Scholar Google	152	4	78
Web of Science	75	9	4
TOTAL	261	19	90

During the last decade, some reviews on Health Care Systems or on their components were published. In order to acquaint the reader with the contents covered in those contributions, a brief summary is presented. Jack and Powers [33] published a review of the literature on demand management, capacity management and performance, covering the period 1986-2006 and extending the review of Smith-Daniels et al [46]. Brailsford and Vissers [11] present the development of the European Working Group ''Operational Research Applied to Health Services'' (ORAHS) as a platform for OR in health, analyzing papers presented at ORAHS meetings since its origin, in 1975. European healthcare systems are the focus of this review and a two-dimension framework analysis is proposed. Developments in applying OR approaches and techniques to healthcare are described and an overview of the main application areas and challenges is presented. Royston [44] presents his personal and selective overview on OR applied to health in the UK, based on his three decades of work in this area. Developments in health and care and in OR contributions to these improvements since the creation of the National Health at UK are described; a reflection on which aspects OR analysts working in health should consider is presented; probable future developments in health and care and subsequent challenges for OR, in order to continue making a significant impact on health care are discussed. Cardoen et al. [15] present a very thorough and meticulous review of recent OR research on operating room planning and scheduling. Works on several fields that are related either to the studied problem or to the technical features are analyzed; the obtained information is structured and pooled in such a way that contributions can easily be linked to each other and compared on multiple aspects. Significant trends in research on operating room planning and scheduling are summarized, and issues and areas currently not well covered and examined are identified. Günal and Pidd [28] published a review on discrete event simulation for performance modelling in health care. They focus on the most recent publications on the topic, especially on performance modelling within hospitals; the papers are classified according to the areas of application that appear in the literature, being the apparent lack of generality and the motifs why specificity dominates discussed. They concluded that most of the reported studies are unit and /or facility specific and also that health-care simulation applications that appear in the literature attempt to provide support for better operational decision-making and planning. Hulshof et al. [29] provide an overview of the typical decisions to be made in resource capacity planning and control in health care, and a review of relevant articles from the field of Operations Research and Management Sciences for each planning decision. For each planning and control decision, not only the key papers but also the methods and techniques applied in the literature to support decision making are reviewed. Considered futures for the use of simulation as a problem solving technique within healthcare settings are proposed by Eldabi et al. [23]. Using a synthesis of trends identified by a selection of experts in the field, academics and industrialists, critical analysis is applied to find the differences between what exists and what could be created based on outlining some major themes. A preference for whole system approaches with more joined up modelling or mixed methods to tackle problems rather than single solution-based practices was observed. Options for the use of simulation within the healthcare domain are presented.

As previously mentioned, all the works contained in "our" database are classified in accordance with the key elements described in Section 2. Table 2 contains information on the percentage of the most cited categories of each one of the considered key elements, for the 90 manuscripts under analysis.

From Table 2, it can be seen that the most cited *entities* are *patients* and *nurses* (as for example in the works [2,3,7,11,13,16,17,18,21,22,30,32,34,35,37,38,39,43,45,47,48,49,50,52,55,56] and [3,32,49, 55], respectively), while *admission, management* and *planning* are the three most frequent *activities* (in particular [2,11,12,13,17,21,26,30,31,34,41,43,45,48,52], [2,7,11,12,13,29,30,31,37,39,45,47,48, 55] and [3,9,11,12,18,25,26,29,35,39,43,45,49,53,55,56], respectively); *beds* are most referenced *resources* (like in [2,3,7,9,12,13,18,21,30,34,35,39,41,48,55]), while *emergency* and *surgery* are the two most used *services* (for instance in [7,9,13,22,30,34,37,45,48] and [16,30,34,37,38,43,47,53,56], respectively). Most of the *planning decisions* are *geographic* and either *local, regional* or *national* (as illustrated in [26,30], [9,48] and [2,37,55], respectively), while *general, acute* and *teaching* hospitals are the most frequent *types of hospital* (as in [12,26,37,48], [7,37,41,48] and [12], respectively). *Inpatient* and *elective* are the two most cited *types of patients* (as in [2,9,12,13,25,30,35,37,45,48] and [7,9,25,30,31,34,37,38,43,52,56], respectively). The most mentioned *performance measures* are *care, capacity, flow* and *efficiency/performance* (as the case of [2,7,9,11,12,13,18,22,25,26,29,30,31,32,38, 39,45,47,48,52], [3,12,13,16,18,22,26,30,31,32,34,35,37,47,48] and [3,7,11,12,16,21,22,25,31,32,38, 39,48,50,52,53,55,56], respectively) being *decision, computational* and *discrete* the most used *models* in the reviewed literature (like in [7,30,31,45,48], [2,26,45,48] and [16,18,26,32,38,47,56], respectively). Most of the mathematical technics quoted are under the scope of *simulation*, *optimization* and *mathematical programming*, which represent 38%, 34% and 26% of the sample, correspondingly (for example, as in [9,11,16,18,22,32,34,35,38,41,47,49,50,53,56], [2,11,18,25,26,29, 30,37,41,43,45,48,49,53,55] and [3,4,7,11,12,13,17,21,29,31,37,39,41,52], respectively), being the tests performed with real data in 22% of the cases (like in [12,16,30,45,48,49,55]).

We also observe that, for every ten analyzed articles, one can found approximately forty eight references on *performance measures*, twenty three on *activities*, fifteen on *mathematical technics*, thirteen on *entities*, eleven on *resources*, nine on *services*, seven on the *type of patients* and on the *type of models*, three on *planning level* and on the *possible applicability* and two on *hospital type*.

Table 2. The three most cited categories of each key element

Key elements	**Percentages of the categories**		
Entities	Patient 62%	Nurses 12%	Doctors 6%
Activities	Admission 44%	Management 39%	Planning 37%
Resources	Beds 40%	Units 11%	Financial 10%
Services	Emergency 32%	Surgery 20%	Intensive care 11%
Planning Level	Local 9%	Regional 7%	National 6%
Hospital Type	General 8%	Acute 7%	Teaching 4%
Patient Type	Inpatient 30%	Elective 21%	Booked 9%
Performance	Care 46%	Flow 41%	Efficiency/performance 40%
Model Type	Decision 17%	Computational 14%	Discrete 13%
Mathematic Technics	Simulation 38%	Optimization 34%	Mathematical Programming 26%
Theory vs. Practice	Real data 22%	Just theoretic 8%	Also implemented 6%

Our literature examination revealed many publications in OR applied to HAS during the last decade. Despite the space constraints, a brief synthesis of some of the analyzed works on OR technics applied to HAS, published in prestigious journals, will follow.

In [12], queuing theory is used to quantify the impact of targeting occupancy rate for all wards at 85%. A decision support system, based on the Erlang loss model, is used and validated. To improve hospital resource efficiency and health care service level, in [39], a multilevel integrative approach based on mathematical programming modeling and simulation analysis is proposed to the planning problem. The performance evaluation is carried out at the operational stage through simulation analysis, and a few effective operational policies are suggested and analyzed to enhance the trade-offs between resource efficiency and service level. In [13], Operations Research technics are used to describe the complexity and dynamics of emergency care chain of cardiac in-patient flow. The optimal bed allocation over the care chain, given a maximum number of refused admissions, is determined and the relation between natural variation in arrivals and length of stay and occupancy rates is analyzed, concluding that both variability in length of stay and fluctuations in arrivals result in large workload variations. Having in mind that managing hospital bed planning is difficult, due to the demand increase and restricted hospital resources, a decision support tool based on an integer linear program for hospital bed planning is proposed in [7]. Elective and acute classes of patients are considered as well as several other constraints, namely incompatibility between pathologies, no mixed-sex rooms and continuity of care, are taken into account. A patient admission scheduling hybrid tabu search algorithm, that supports operational decisions in a hospital and automatically assigns patients to beds, is described in [21]. The algorithm not only takes into account the medical needs of the patients and their preferences, but it also keeps the number of patients in the different departments balanced. The performance of the algorithm is compared with an integer programming approach using a randomly generated data set. In [52], the impact on welfare implications of a 0-1 linear programming model to solve the operating room planning problem, taking a patient perspective, is evaluated. For a given planning period, the model determines the allocation of the operating room block times to surgical sub-specialties, i.e. the so called Master Surgical Schedule Problem, together with the subsets of elective patients to be operated on in each block time, i.e. the so called Surgical Case Assignment Problem. The model performance is verified by applying it to a real scenario; the computational results obtained indicate that the model can solve all test problems. A multi-neighborhood local search procedure to solve a healthcare problem, known as the Patient Admission Scheduling problem, is proposed in [17]. Local search is performed, lower bounds are computed and the application to the dynamic case, in which admission and discharged are not predictable in advance is proposed. Favorable results are obtained. Tactical plans allocate available resources to various care pathways and determine the selection of patients, on a particular stage of their care pathways, to be served; in [31], a method to develop a tactical resource allocation and an elective patient admission plan are proposed. A Mixed Integer Linear Programming framework is used and involves the cooperation with multiple resources, multiple time periods and multiple patient groups with various uncertain care pathways, thereby integrating decision making for a chain of hospital resources. Computational results indicate that more equitable distribution of the resources and control of patient access times, of the number of patients served and of the fraction of allocated resource capacity are obtained. A three-phase, hierarchical approach for the weekly scheduling of operating rooms is developed in [53]. The aim is to suggest an integrated way of facing surgical activity planning in order to improve overall operating theatre efficiency, in terms of overtime and throughput as well as waiting list reduction, while improving department organization; a bin packing-like problem, the blocked booking method and simulation software are used. A model for the surgery admission planning problem and a meta-heuristic algorithm for solving it are presented in [43]. The algorithm uses simple Relocate and Two-Exchange neighborhoods, governed by an iterated local search framework. The problem's search space associated with these move operators is analyzed in terms of patient waiting time, surgeon overtime, and waiting time for children in the morning on the day of surgery. Computational results for a set of benchmarks, based on the admission planning problem in a chosen hospital, are presented. Different categories of cardiothoracic surgery patients can be distinguished based on their requirement of resources; in [3], the planning problem is considered at the tactical level to generate a master surgical schedule that realizes a given target of patient throughput and optimizes an objective function for the utilization of resources. The problem was mathematically approached using mixed integer linear programming with a stochastic length of stay. Good performance on target utilization levels was obtained. With the aid of a pre-existing data-integrated simulation model to evaluate nurse-patient assignments, two policies to make nurse-to patient assignments, when new patients are admitted during a shift, are developed in [49]; in one of them, a heuristic policy assigns a newly-admitted patient to the nurse who has performed the least assigned direct care among all the nurses and, in the other, a partially-optimized policy seeks to minimize the difference in workload among nurses for the entire shift by estimating the assigned direct care. Results comparing the two policies are presented. A new method, simple to use as the ratio method while minimizing the roundabout approaches of these methods, is proposed in [41] to determine the number of acute hospital beds. The simulation of an unusual increase in admission requests had no consequence on the bed number selected, which indicates the robustness of the method; software was developed and is available for use. In [18], a multi-stage stochastic methodology to balance inpatient bed unit utilizations in a hospital is proposed. Queuing network analysis and optimization are used to achieve balanced targets of bed unit utilization while building hospital staff involvement. Discrete event simulation is, then, used to maximize flow through the system including nonhomogeneous effects of daily and hourly peak loading, nonexponential lengths of stay, and blocking behavior. A major hospital is analyzed with the methodology and results being validated against field data. A nonhomogeneous discrete time Markov model for modeling a patient flow through the care system is presented in [26]. This model can be used for admission scheduling, resource requirement forecasting and resource allocation, in order to satisfy the demand or resource constraints or to meet the expansion or contraction plans in a hospital and community based integrated care system. It is also a useful tool for care managers and policy makers who need to make strategic management decisions. An application of the model to an elderly care system is described. In order to create an expedited patient care queue for admission to a hospital, an optimal admission threshold policy using controls on the scheduled and expedited gateways for a new Markov decision process model is characterized in [30]. A practical policy based on insight from the analytical model that yields reduced emergency blockages, cancelations, and off-unit census via simulation based on historical hospital data is presented. The lack of adequate numbers of hospital beds to accommodate the injured is a major problem in public hospitals; a dynamic system that announces the status of each bed when it changes with admission or discharge of a patient is proposed in [2] and implement national-wide to control the occupancy of beds, in order to manage cross hospital referral patient in crowded hospitals that do not have enough beds for admission patient. A national wide network for bed management, that helps to distribute injured patient in the hospitals, is provided. In [55], Markov methods, Petri net and genetic algorithms are used; development of models and the recourse to intelligent optimization for solving problems, such as improve work force efficiency and the percentage of bed occupation and also regulate shift work program, are described; cost reduction and work force planning are achieved.

4. Conclusions

In this study, we analyzed the literature on Health Admission Systems over the period 2005-2014, mainly through works published on books, journals, proceedings, as well as on MSc and Ph.D. thesis and working papers. It is important to note that, throughout this decade, an increase in the number of publications under this scope is observed, which shows the growing interest of researchers in this area.

The hospital, as provider of services for patients, has long been recognized as an obvious subject for OR Methods. Most studies focus on variability, prediction, flow and census control. The stochastic models have been used mostly to study census variability and prediction. The optimization models and mathematical programming algorithms have been applied chiefly for designing admission policies and to exercise control over the census and the flow. Complex problems which may not be treated analytically have been studied experimentally through the use of simulation. However, only a little more than a fifth of the analyzed works are based on real data and less than 6% of them were implemented in practice, which shows that there is still much to be done in this area.

Finally, we stress out that our intention was to analyze the main characteristics of the research in OR applied to HAS, an ambitious task interacting with many other fields of research, namely several health fields, economics and management. We also observe that the authors continue to study this issue extending the time horizon under consideration.

5. References

[1] American Hospital Association 1961 *Hill-Burton Act P L 79-725 (1946). Area-wide planning for hospitals and related health facilities* (Washington DC: Superintendent of Documents, U.S. Government Printing Ofllco)

[2] Abedian S, Kazemi H, Riazi H and Bitaraf E 2013 Cross hospital bed management system *Stud Health Tech Informat* **205** 126-30

[3] Adan I, Bekkers J, Dellaert N, Vissers J and Yu X 2009 Patient mix optimisation and stochastic resource requirements: A case study in cardiothoracic surgery planning *J Health Care Manage Sci* **12**(2) 129-41
[4] Ayvaz N 2010 Allocation of hospital capacity to multiple types of patients, *JRPM* **9**(5) 386–98
[5] Bailey N T J 1954 Queuing for medical care *Appl Stat* **3** 137-45
[6] Balintfy J L 1960 A stochastic model for the analysis and prediction of admission and discharges in hospitals *Management Sciences, Models and Techniques: Proc of XXXVI ORAHS Conference* vol 2 eds. C.W. Churchman and M. Verhulst (New York: Pergamon) pp 288-99
[7] Ben Bachouch R, Guinet A, and Hajri-Gabouj S 2012 An integer linear model for hospital bed planning *Int J Prod Econ* **140**(2) 833-43
[8] Boldy D 1976 A review of the application of mathematical programming to tactical and strategic health and social service problems *J. Oper. Res. Soc.* **27**(2) 439-48
[9] Bowers J and Mould G 2005 Ambulatory care and orthopaedic capacity planning *J Health Care Manage Sci* **8**(1) 41-7
[10] Blumberg M S 1961 DPF concept helps predict bed needs *Mod Hosp* **97** 75-81
[11] Brailsford S and Vissers J 2011 OR in healthcare: A European perspective *Eur J Oper Res* **212** 223–34
[12] Bruin A M, Koole G M, Zanten L V and Bekker R 2010 Dimensioning hospital wards using the Erlang loss model *Ann Oper Res* **178**(1) 23-43
[13] De Bruin A M, Van Rossum A C, Visser M C and Koole G M 2007 Modeling the emergency cardiac in-patient flow: an application of queuing theory *J Health Care Manage Sci* **10**(2) 125-37
[14] Campbell W B et al 2002 A 25-year study of emergency surgical admissions *Ann R Coll Surg Engl* **84**(4) 273–77
[15] Cardoen B, Erick D and Jeroen B 2010 Operating room planning and scheduling: A literature review *Eur J Oper Res* **8** 101-4
[16] Cardoen B and Demeulemeester E 2008 Capacity of clinical pathways — a strategic multi-level evaluation tool *J Med Syst* **32**(6) 443-52
[17] Ceschia S and Schaerf A 2011 Local search and lower bounds for the patient admission scheduling problem *Comput Oper Res* **38**(10) 1452-63
[18] Cochran J K and Bharti A 2006 A multi-stage stochastic methodology for whole hospital bed planning under peak loading *IJISE* **1**(1) 8-36
[19] Connor R J 1960 *A hospital inpatient classification system* (Johns Hopkins University: Doctoral dissertation)
[20] De Oliveira M J F 1982 *The use of information in Planning hospital admission with special reference to Glasgow Western Infirmary 1977-79* (Strathclyde University: Doctoral dissertation)
[21] Demeester P, Souffriau W, De Causmaecker P and Vanden Berghe G 2010 A hybrid tabu search algorithm for automatically assigning patients to beds *Artif Intell Med* ***48***(1) 61-70
[22] Diefenbach M and Kozan E 2008 Hospital Emergency Department simulation for resource analysis *Ind Eng Manag Syst* *7*(2), 133-42
[23] Eldabi T, Paul RJ and Young T 2007 Simulation modeling in health care: Reviewing legacies and investigating future *J. Oper. Res. Soc.* **9** 262-70
[24] Fries BE 1976 Bibliography of operational research in health care systems *Opns Res* **24** 801-14
[25] Gallivan S and Utley M 2005 Modelling admissions booking of elective in-patients into a treatment centre *IMA J Management Math* **16**(3) 305-315
[26] Garg L 2010 A non-homogeneous discrete time Markov model for admission scheduling and resource planning in a cost or capacity constrained healthcare system *J Health Care Manage Sci* **13**(2) 155-69

[27] Gonseth J, Guallar-Castillón P, Banegas J R, and Rodríguez-Artalejo F 2004 The effectiveness of disease management programmes in reducing hospital re-admission in older patients with heart failure: a systematic review and meta-analysis of published reports *Eur. Heart J.* **25** 1570–1595

[28] Günal M M and Pidd M 2010 Discrete event simulation for performance modelling in health care: a review of the literature. *J Simul* **4**(1) 42-51

[29] Hans E W, Kortbeek N, Hulshof P J H, Bakker P J M and Boucherie R J 2012 Taxonomic classification of planning decisions in health care: a structured review of the state of the art in OR/MS *Health Syst* **1** 129–75

[30] Helm J E, AhmadBeygi S and Oyen M P V 2011 Design and analysis of hospital admission control for operational effectiveness *Prod. Oper. Manag.* **20**(3) 359-74

[31] Hulshof P J, Boucherie R J, Hans E W and Hurink J L 2013 Tactical resource allocation and elective patient admission planning in care processes. *J Health Care Manage Sci* **16** 152–66

[32] Ignone G, Mossa G, Mummolo G, Pilolli R and Ranieri L 2013 Increasing public healthcare network performance by de-hospitalization: A patient pathway perspective *SOIJ* **6**(1) 85-107

[33] Jack E P and Powers T L 2009 A review and synthesis of demand management, capacity management and performance in health-care services *Int. J. Manag. Rev.* **11**(2) 149-74

[34] Kolker A 2009 Process Modeling of ICU Patient Flow: Effect of Daily Load Leveling of Elective Surgeries on ICU Diversion *J Med Syst* **33**(1) 27–40

[35] Kumar S (2011) Modeling patient flow operation of a US urban county hospital *Technol Health Care* **19**(4) 247-60

[36] Lagergren M 1998 What is the role and contribution of models to management and research in the health services? *Eur J Oper Res* **105**(2) 257-66

[37] Lane D C, Husemann E 2008 System dynamics mapping of acute patient flows *J. Oper. Res. Soc.* **59**(2) 213–24

[38] Levy A R, Sobolev B G, FitzGerald J M, Kuramoto L, Schechter M and Sanchez V 2008 Evaluation of booking systems for elective surgery using simulation experiments *Healthc Policy* **3**(4) 113–24

[39] Ma G and Demeulemeester E 2013 A multilevel integrative approach to hospital case mix and capacity planning *Comput Oper Res* **40**(9) 2198-207

[40] Milsun J H, Turban E and Vertinsky I 1973 Hospital admission systems: Their evaluation and management *Manag. Sci.* **19** 646-66

[41] Nguyen J M, Six P, Antonioli D, Glemain P, Potel G, Lombrail P and Le Beux P 2005 A simple method to optimize hospital beds capacity *Int J Med Inform* **74**(1) 39-49

[42] Nuffield Provincial Hospital Trust 1960 *Studies in the functions and design of hospitals: the report of an investigation* (London: Oxford University Press)

[43] Riise A and Burke E K 2011 Local search for the surgery admission planning problem *J Heuristics* **17**(4) 389-414

[44] Royston G 2009 One hundred of operational research in health. *J. Oper. Res. Soc.* **60** 169-79

[45] Sharieff G Q, Cantonis M, Tressler M, Whitehead M, Russe J and Lovell 2014 Decreasing Avoidable Hospital Admissions With the Implementation of an Emergency Department Case Management Program *Am J Med Qual* **29**(3) 200-05

[46] Smith-Daniels V L, Schweikhart S B and Smith- Daniels D E 1988 Capacity management in health care services: Review and future research directions *Dec Sci* **19** 889-918

[47] Sobolev B, Harel D, Vasilakis C and Levy A 2008 Using the statecharts paradigm for simulation of patient flow in surgical care *J Health Care Manage Sci* ***11***(1) 79-86

[48] Subbe C P, Kellett J, Whitaker C J, Jishi F, White A, Price S, Ward-Jones J, Hubbard R E, Eeles E, Williams L 2014 A pragmatic triage system to reduce length of stay in medical emergency admission: Feasibility study and health economic analysis *Eur J Intern Med* **25**(9) 815–20

[49] Sundaramoorthi D, Chen V C, Rosenberger J M, Kim S B and Buckley-Behan D F 2010 A data-integrated simulation-based optimization for assigning nurses to patient admissions *J Health Care Manage Sci* **13**(3) 210-21
[50] Tai G, Williams P and Lei Y 2010 Simulation based analysis of patient arrival to health care systems and evaluation of an operations improvement scheme *Ann Oper Res* **178**(1) 263-79
[51] Tam V C, Knowles S R, Cornish P L, Fine N, Marchesano R and Etchells E E 2005 Frequency, type and clinical importance of medication history errors at admission to hospital: a systematic review. *J Ayub Med Coll Abbottabad* **173** 510-515
[52] Testi A and Tànfani E 2009 Tactical and operational decisions for operating room planning: Efficiency and welfare implications *J Health Care Manage Sci* **12**(4) 363-73
[53] Testi A, Tànfani E and Torre G 2007 A three-phase approach for operating theatre schedules *J Health Care Manage Sci* **10**(2) 163–72
[54] Thompson JD and Fetter RB 1969 Economics of occupancy with varying mixes of private and other patient accommodations: A simulation. *Health Serv Res* **4** 42-52
[55] Vaezi M and Darvish N 2011 Modeling and scheduling intelligent method's application in increasing hospitals' efficiency *J Theor Appl Inf Technol* **14**(9) 34
[56] VanBerkel P T and Blake J T 2007 A comprehensive simulation for wait time reduction and capacity planning applied in general surgery *J Health Care Manage Sci* **10**(4) 373-85
[57] Young J P 1962 *A queuing theory approach to the control of Hospital Inpatient census system* (Johns Hopkins University: Doctoral dissertation)
[58] Wing K R 1982 The Community Service Obligation of Hill-Burton Health Facilities *B.C.L. Rev.* **23**(3) 577

Acknowledgments
The authors thank the University of Trás-os-Montes and Alto Douro and the Technology Center of the Federal University of Rio de Janeiro. They also thank the CIO–Operations Research Center that is supporting this work using funds from FCT-Fundação para a Ciência e a Tecnologia, under the project PEst-OE/MAT/UI0152.

Journal of Physics: Conference Series **616** (2015) 012010 doi:10.1088/1742-6596/616/1/012010

Occupational health and safety: Designing and building with MACBETH a value risk-matrix for evaluating health and safety risks

D F Lopes[1], M D Oliveira and C A Bana e Costa

CEG-IST, Centre for Management Studies of Instituto Superior Técnico, Universidade de Lisboa, Avenida Rovisco Pais, 1049-001 Lisbon, Portugal

[1]E-mail: diana.lopes@tecnico.ulisboa.pt

Abstract. Risk matrices (RMs) are commonly used to evaluate health and safety risks. Nonetheless, they violate some theoretical principles that compromise their feasibility and use. This study describes how multiple criteria decision analysis methods have been used to improve the design and the deployment of RMs to evaluate health and safety risks at the Occupational Health and Safety Unit (OHSU) of the Regional Health Administration of Lisbon and Tagus Valley. 'Value risk-matrices' (VRMs) are built with the MACBETH approach in four modelling steps: a) structuring risk impacts, involving the construction of descriptors of impact that link risk events with health impacts and are informed by scientific evidence; b) generating a value measurement scale of risk impacts, by applying the MACBETH-Choquet procedure; c) building a system for eliciting subjective probabilities that makes use of a numerical probability scale that was constructed with MACBETH qualitative judgments on likelihood; d) and defining a classification colouring scheme for the VRM. A VRM built with OHSU members was implemented in a decision support system which will be used by OHSU members to evaluate health and safety risks and to identify risk mitigation actions.

1. Introduction

Occupational health and safety (OH&S) is "the science of the anticipation, recognition, evaluation and control of hazards arising in or from the workplace that could impair the health and well-being of workers" [1] (p. viii). Proper management of OH&S risks has been recognized as a social concern topic and as a workers' right by many organizations, including the International Labour Organization, the World Health Organization, the European Commission and national governments, in line with promoting a healthy and productive labour force [2].

A lot of research has been devoted to the development of tools for risk assessment, with risk matrices (RMs) being one of the tools most widely used to evaluate risks in general, and OH&S risks in particular [3-5], as they are easy to handle, demand for limited expertise, and are enabled by several popular software packages. Nonetheless, several studies point out that RMs violate theoretical principles that compromise their feasibility and use [6-9].

The Occupational Health and Safety Unit (OHSU) of the Regional Health Administration of Lisbon and Tagus Valley (RHA LVT) is responsible for evaluating OH&S risks for all the individuals working in primary care centres and in administrative offices of the RHA LVT. Although OHSU team members have been using RMs for OH&S risk assessment [5], they have recognized several difficulties in their use and asked for our help to improve their risk evaluation process.

This study reports how tools from Multiple Criteria and Decision Analysis (MCDA) have been used to design a Value Risk-Matrix (VRM) to be used by the OHSU to evaluate OH&S risks, as well as briefly describes a decision support system (DSS) that was created to enable the use of that VRM. Several components of the development of VRMs make a contribution to literature, for instance, the modelling of risk impact dimensions with the MACBETH-Choquet procedure (the first application of this procedure) and the modelling of probabilities with MACBETH.

This article is organized in five sections. Section 2 briefly introduces RMs, their problems and challenges. Section 3 describes what a VRM is, and the activities required for its design at the OHSU. Section 4 briefly describes how each activity was applied with OHSU members and the DSS that was created to assist OHSU. Section 5 presents concluding remarks.

2. Literature review

In the European Union several people die per year as a consequence of occupational accidents and diseases, with most of these being avoidable [10]. The first step in preventing them is risk assessment [10], which provides decision makers (DMs) with an improved understanding of risks that could threaten their goals and on the extent to which mitigation actions need to be implemented [11].

Research and regulation of OH&S are a relatively recent phenomenon [12]. Within common frameworks of management of OH&S risks, several guidelines [4], research studies [13-15] and international standards [11] advocate the use of RMs. RMs are used and recommended to evaluate and mitigate risks in many health care contexts, such as by the National Patient Safety Agency from the United Kingdom [4]. In this tool, individual sources of risk are typically characterized in terms of risk probability and impact. Table 1 illustrates the RM used by the OHSU of the RHA LVT to evaluate OH&S risks [5], which is similar to many other RMs. That RM includes two 4-level rating scales in each probability and impact component. The risk score is given by crossing both information, i.e. by multiplying its probability and impact. This is complemented by information on risk criticality, as depicted by the traffic light system that colors each combination of probability and impact in the RM (corresponding to the 5 risk categories "very low", "low", "moderate", "high" and "very high") [5].

Table 1. Example of a quantitative RM (adapted from [5]), with five risk categories (from light grey to dark grey) "very low", "low", "moderate", "high" and "very high".

		Severity of the injury			
		1. Very low	2. Low	3. High	4. Very high
Probability of the event	1. Unlikely	1. Very low	2. Very low	3. Low	4. Moderate
	2. Likely	2. Very low	4. Moderate	6. Moderate	8. High
	3. Quite likely	3. Low	6. Moderate	9. High	12. Very high
	4. Very likely	4. Moderate	8. High	12. Very high	16. Very high

Despite their wide use, RMs violate some theoretical principles [6-8] that compromise their feasibility and use, such as (1) misuse of rating scales for probability and impact which leads to quantitatively meaningless ratings, (2) disregard of cumulative effects of multiple impacts and (3) improper selection and rating of risk mitigation actions based on inefficient allocation of resources (for further details, consult [6-9]).

Besides these problems that have been reported in literature, judgmental dependencies between risk impacts in different dimensions may occur, and therefore modelling should identify and account for these value interdependencies (note that value interdependencies differ from cases of statistical or physical correlations, as they depend on the DM perspectives and values). Situations of preference interdependencies have been found in multiple contexts, and have been recognised as a major challenge in MCDA literature [16], with several studies exploring the use of Choquet integral (CI) operators to model these interdependencies, as these operators demand for weaker conditions in comparison to other models, such as the multilinear one that demands for the respect of weak difference independence conditions [17]. There is thus scope for improving tools to assist the evaluation of OH&S risks, using the same framework as RMs, but overcoming the methodological issues above identified and addressing the challenge of identifying and modelling value interdependencies in risk impacts.

3. Methodological approach

Our work with the OHSU started by a set of interviews that identified the difficulties felt by members of the unit in the process of evaluating OH&S risks, and their perceived challenges. Summarily, these interviews have shown: difficulties in identification and measurement of risk impacts due to the lack of technical instruments (e.g., for measuring noise levels), as well as linking risk sources with their consequences due lack of access to systematic evidence; great difficulties in assessing probabilities given a lack of a registry and of a summary of evidence in the area; and disagreement with some of the values implicit in the RM in use (e.g., third-degree burns and a death are similarly rated).

In order to correct problems from RMs, we propose using a VRM – as shown in Figure 1, a VRM is similar to the RM as it combines the two risk components, impact and probability, to define a few risk categories (e.g., from grade I to grade V). However, (discrete) risk impacts present in common RMs – as in Table 1 – are replaced by (continuous) multicriteria value scores, (discrete) probabilities by a (continuous) probability scale by adopting specific MCDA tools, such that discontinuities between adjacent categories (as in Table 1) are avoided and the principle of translation is respected. The value and probability scores of a risk determine its position in the VRM and users can easily observe whether that risk is seen as critical to the organization and may require specific managerial attention [9].

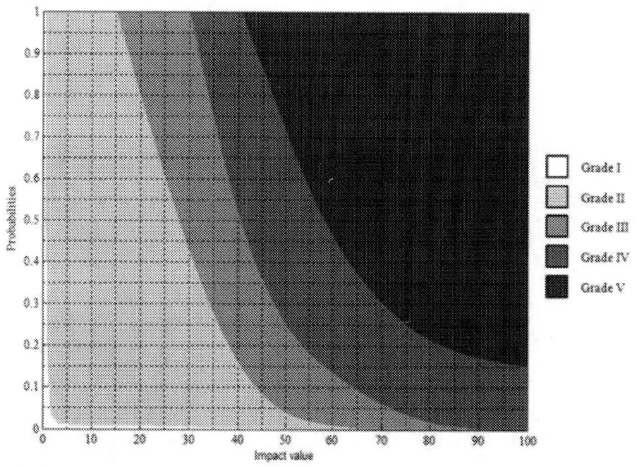

Figure 1. VRM with five categories, from grade I to grade V, built to assist the OHSU on evaluating OH&S risks.

Building the VRM to the OHSU followed a socio-technical approach and a sequence of several interconnected activities, as depicted in Figure 2. The social component consisted in the development of these activities within various meetings, workshops and decision conferences with the DMs, which in our case were the professional members of the OHSU with different backgrounds and distinct viewpoints regarding the evaluation of OH&S risks (namely, doctors, nurses, a sanitary engineer and OH&S risk technicians). The social component enabled the development of a requisite VRM (with model requisiteness as defined by [18]) that captured the perspectives of the team members. On the technical side, developing the VRM included: (1) the structuring of risk impacts, (2) the construction of the value scale for risk impacts with MACBETH-Choquet procedure, (3) the construction of a system of rules for defining subjective probabilities (using reference probabilities constructed with the MACBETH approach), (4) the design of risk categories, and (5) building a DSS to enable the use of the VRM for OH&S risk assessment. An explanation on what consisted each of these activities, and how it was developed with the DMs will be provided in the next section. We have used the MACBETH approach to build the VRM as it is recognised as being a simple and transparent MCDA tool in modelling complex multidimensional problems, based only on non-numerical judgments about differences in attractiveness between evaluation elements, avoiding the "difficulty" [17] of expressing value judgments numerically [9]. Although the MACBETH approach has been previously used in real contexts, this is the first time that it was applied the MACBETH-Choquet procedure and that MACBETH was adapted and applied for probability assessment (requiring a change in its formulation and questioning protocol).

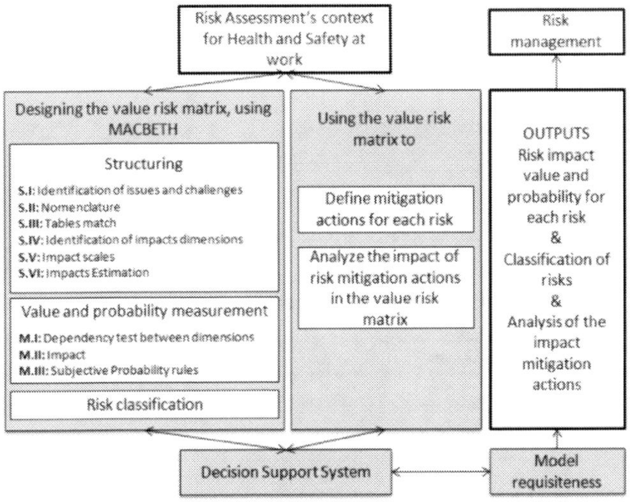

Figure 2. Activities developed for building the VRM and the DSS.

4. Steps in the design and deployment of the Value Risk-Matrix

4.1 Structuring the multicriteria model of risk impacts

Structuring a multicriteria model of risk impacts typically consists of an interactive learning process with the DM group including: (a) the specification of the relevant dimensions for appraising risk impacts (following the approach of structuring "value dimensions" of von Winterfeldt et al. (1986)); (b) the assignment of a descriptor of impact in each risk dimension. In our case, departing from the risk evaluation challenges above identified, the structuring phase was done through the following activities: (S.I) definition of a clear nomenclature, (S.II) construction of tables that link risk sources to consequences in health, based upon evidence in the area (S.III) specification of the relevant dimensions for appraising risk impacts; (S.IV) construction of the impact scales in each dimension and (S.V) estimation of the impact on each dimension, according to evidence in the area.

Step S.I: Nomenclature

OHSU's guidelines for using RMs and all the reports based on it were analyzed. This allowed us to detect incoherencies and inconsistencies, namely on the nomenclature in use, and given the ambiguity in several risk terms, the terms in use were discussed and defined with members of the OHSU. Accordingly, the risk terms defined in Table 2 were adopted.

Step S.II: How to identify and measure risk sources and their consequences?

Given the difficulties reported by the DMs on identifying and measuring the expression levels of risk sources and their consequences due to the lack of technical tools (e.g. for measuring noise levels), we made an intensive literature review to build qualitative (or quantitative or pictorial) descriptors of impact that link different risk sources and expression levels with health consequences. Table 3 exemplifies the descriptor for the risk source "noise", which links, for example, the noise level present in a waiting room with quantitative levels, according to literature (for instance, making explicit the exposure limit imposed by legislation, 87 dB [19]). Descriptors were similarly built for other risk sources.

Table 2. Nomenclature proposed and validated by the DMs.

Terms	Description
Event	Situation or phenomenon in which there is uncertainty in its occurrence.
Risk source (or hazard)	Source of potential adverse effects or a situation capable of causing adverse effects in terms of health (injuries or illnesses) [7].
Expression levels of the risk source	A risk source can be expressed in different ways that we called expression levels.
Measure of the expression levels related to a risk source	Identification and awareness/measurement of qualitative or quantitative or pictorial expression levels.
Risk	"A risk is a random event that may possibly occur and, if it did occur, would have a negative impact on the goals of the organization" [20] (p.3). Typically, it is characterized by two components: its probability of occurrence and impacts [10].
Risk typology	Risk nature or risk classification in five types enshrined in the Decree-Law no. 441/91 and in the Law no. 99/2003.
Consequence	Outcome of a risk [11] (bodily injury, functional disorder or disease) affecting the objectives of an organization.
Impact	Severity [20] or effects of the consequences' occurrence.

Table 3. Descriptor of impacts of the risk source "noise" [21, 22].

Description of noise	dB	Health consequences
Normal breathing - Air conditioning	10 – 50	Without effect
Normal conversation (public lounges, offices, cafes, bars) - Aspirator	60-70	Annoyance pronounced
Hairdryer	80	Annoyance and possible increase in the hearing threshold level
Backhoe	85	Possible increase in the hearing threshold level
Exposure Limit	87	Hearing loss of 10-15 dB in a working period from 1 to 2 years
Mowing machine	89	Hearing loss of 50 dB in the working period of 50/52 years
Ambulance siren	120	Pain and hearing loss treatable or not treatable
Jet plane	140	Pain and acoustic trauma. Hearing loss may be accompanied by a buzzing

Step S.III: Identification of impact dimensions

This step consisted on identifying the impact dimensions to evaluate risk impacts. Due to the recursive nature of our proposal methodology, the impact dimensions were iterated until the three dimensions – "employee's health" (EH), "work capability" (WC) and "absenteeism" (A) – were identified, as depicted by the value tree in Figure 3. The DMs were concerned about OH&S risk impacts for the health of the employee and for the employers (if the employee returns, and with which capabilities).

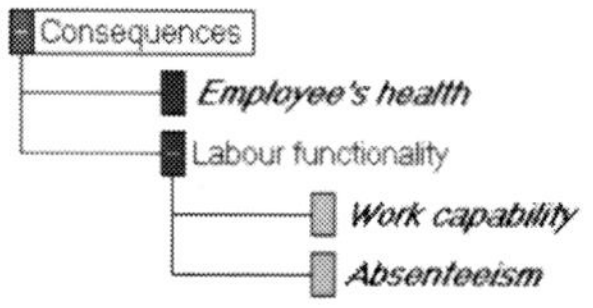

Figure 3. Final value tree built in decision conferences with the DMs (three key dimensions in italic).

Step S.IV: Impact scales construction

This step consisted in building impact scales for each dimension. For the "employee's health" dimension, the DMs were concerned about the impact on the employee's health. This led to the construction of a continuous impact scale based on the disability adjusted life year measure, which varies from the *status quo* (corresponding to no impact, i.e., 0 Years of Healthy Life Lost (YHLL)) to the worst level (corresponding to the death of the employee that implies 34 YHLLs). For the "work capability" dimension, a 5-level qualitative impact scale was built by adapting the legal nomenclature [23], namely: "null disability", "recoverable disability", "irrecoverable partial disability with return to work", "irrecoverable partial disability with no return to work" and "irrecoverable disability". Regarding the "absenteeism" dimension, a continuous impact was constructed based on the number of days of absence from work.

Step S.V: How to estimate the consequences' impact in each dimension?
Given the DMs difficulty in estimating impacts of consequences due to lack of knowledge, an intensive literature review was done to build tables linking the consequences of different risks to the impact scales. For instance, according to literature, an amputation of a finger was shown to correspond to 3.5 YHLL [24, 25], to an irrecoverable partial disability and to 26 days of absence from work [26].

4.2 Impact value measurement with the MACBETH-Choquet procedure
Building a value measurement scale for risk impacts requires the conversion of impact into value. This activity was carried out by using the MACBETH-Choquet procedure that allows for modelling interdependencies between impact dimensions. This approach combines MACBETH with the CI, through the use of a MACBETH global matrix filled with judgments of global attractiveness between risk impacts, being the basis for determining the CI parameters (for further details, consult [27]). This value measurement technique was chosen as we found out that there was preferential dependence between some pairs of dimensions (for instance, work capability assessment judgmentally depended upon employee's health, according to the DMs' preferences), and this work allowed for using that procedure in a real context for the first time.

The first step for applying the MACBETH-Choquet procedure was to define all the feasible combinations of different impact levels across all the three dimensions – in our example, 36 combinations of impacts were considered. In particular, those combinations vary from the *status quo* corresponding to no impact (0 YHLL, null disability and 0 days of Absence from Work (AW)) to the worst level corresponding to states equivalent to death (34 YHLL, irrecoverable disability and 18 years of AW).

MACBETH-based elicitation was then used with the DMs to quantify the relative attractiveness of reverting each set of impacts through a qualitative pairwise comparison questioning mode, supported by the Microsoft Office PowerPoint 2007 (MOP 07). This questioning protocol was built to be applied in a sequential and interactive manner (using the *ActiveX textbox* tool). To illustrate the protocol, the DMs were asked about the attractiveness of fully reversing (or cancelling out) the following combination of impacts: 34 HLL, irrecoverable disability and 18 years of AW, in the MACBETH categorical scale "null" (C_0), "very weak" (C_1), "weak" (C_2), "moderate" (C_3), "strong" (C_4), "very strong" (C_5) or "extreme" (C_6) – that are represented in the matrix of Figure 4(a) by the respective indices "1" to "6" for the sake of saving space. The DMs answered that reversing this risk impact was extremely attractive, with this being then recorded in the MACBETH matrix of qualitative judgements presented in Figure 4. The same type of questioning protocol was applied for all the other combinations of risk impacts (as explained in [28]) and the answers were recorded in the M-MACBETH software (Figure 4). The DMs revised some judgments when inconsistencies between answers were detected, and after discussion, a global (numerical) score was validated (as shown in Figure 4(b)).

Regarding the CI parameters, the Shapley and interaction parameters are key concepts for the understanding and analysis of interdependencies between dimensions. According to Grabisch (1997), the Shapley value can be defined as an average value of the contribution of dimension i alone considering different baselines in the other dimensions. With regard to the interaction parameter between 2 dimensions for instance, it can be defined as an average of the added value given by putting together the 2 dimensions together, all the baselines being considered. The interaction parameters I_{ij} range in the interval [-1,1] with: (a) $I_{ij} > 0$ for a synergistic behaviour between dimensions *i* and *j*, (b) $I_{ij} < 0$ for an antagonistic behaviour between dimensions *i* and *j*, and (c) $I_{ij} = 0$ when there is no interaction between dimensions *i* and *j*. The mathematical representations of the Shapley and the interaction parameters are given by (1) and (2), respectively.

$$\begin{cases} s_i = \sum_{K \subset M \setminus i} \frac{(m-|K|-1)!|K|!}{m!} [V_{Ag}(B_K \cup B_i) - V_{Ag}(B_K)] \\ \sum_{i=1}^{m} s_i = 1 \end{cases} \tag{1}$$

$$I(A) = \sum_{K \subset M \setminus A} \frac{(m-|K|-|A|)!|K|!}{(m-|A|+1)!} \sum_{L \subset A} (-1)^{|A|-|L|} V_{Ag}(B_K \cup B_L), \quad A \subset M, A \neq \emptyset \tag{2}$$

Using the rescaled value scores (from Figure 4(b)) into Eqs. (1) and (2), we obtained the CI parameters: $s_{EH} = 295/338, s_{WC} = 32/338$ $s_A = 32/338$, $I_{EH\&WC} = 22/169, I_{EH\&A} = 5/169$ and $I_{WC\&A} = I_{EH\&WC\&A} = 0$.

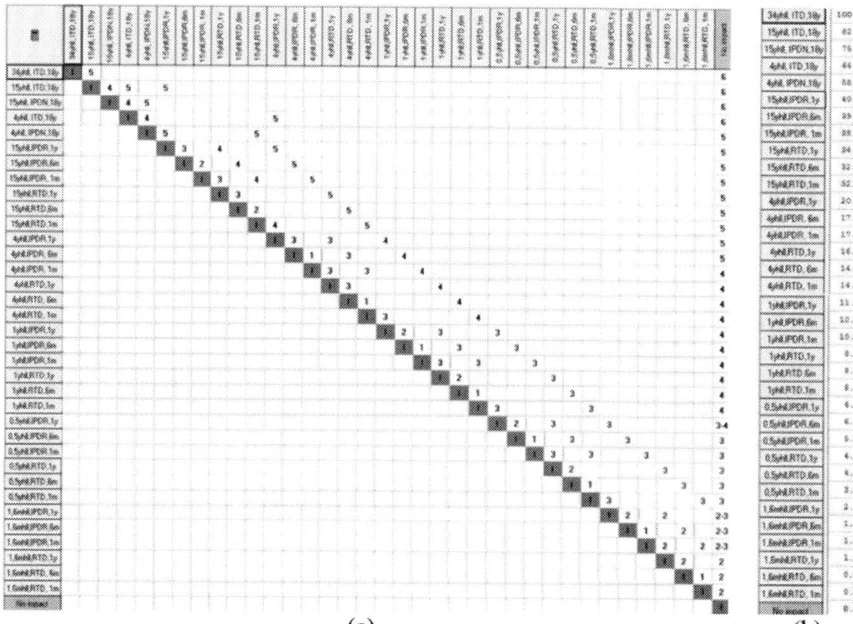

(a) (b)

Figure 4. Global MACBETH matrix of judgments (a) and the corresponding numerical impact scale validated by the OHSU group (b).

4.3 Subjective probabilities through a system of rules

Following difficulties felt by the OHSU group in selecting probabilities for risks (historical information on risks was not available), as well as technical and behavioural inconsistencies in associating numerical probabilities to verbal expressions pointed out in literature [29], we have designed a system to help the group to construct subjective probabilities. This system had as underlying assumptions: probabilities elicitation in a verbal formal [30]; probability can be seen as “a personal judgment of the likelihood of a proposition or event rather than an objective fact" [31], with a constructive view of probabilities [17] being useful in the context.

Building a system to construct subjective probabilities involved: a selection of a wide range of representative risk events that covers all the likelihood scale, according to literature and to experience from the members of the OHSU; the construction of a numerical probability scale for these representative risk events, using the MACBETH approach adapted for probability assessment; and then a system of rules in which, for a new risk, the OHSU will find the most similar representative event, and then will re-adjust the numerical probability according to *pros* and *cons*, i.e. by aspects that can contribute for a higher/lower likelihood (for instance, if employees have attended a course on OH&S risks, the likelihood of some risk events occurring will be lower).

After a set of representative risks was selected, the MACBETH questioning mode was adapted for the probability assessment context, with the OHSU group being asked for verbal judgments about differences of likelihood between risk events. This approach is grounded on (a) the definition of two reference levels, a certain event and an impossible event which correspond to 1 and 0, respectively; (b) a qualitative pairwise comparisons questioning protocol based on questions-type such as "How likely is risk x to occur?", or said in another way "which is the difference in likelihood between x and an impossible event?" (with x being fully specified as a risk event). This probability assessment consisted on a conditional assessment of subjective probabilities, as it required a certain *priori* knowledge [32] about the risk event (for instance, with "effectiveness of security measures already implemented" and "exposure" being made explicit). Additional questions were asked to distinguish between which events were more probable than others (for further details consult [9]). After discussion with the OHSU members, the consistent MACBETH matrix present in Figure 5 was obtained, and the final numerical scale of subjective probabilities present in Figure 5(b) was validated.

Finally, a generic system with specific rules for the assessment of probability of a new risk was designed: for a new risk, the user of the VRM should find a reference event with each to compare the new risk, and then adjust the probability according to the *pros* and *cons* that may increase or decrease probability (that is, specific factors that will contribute for an increase (or decrease) in the probability, such as whether the employee received training to manage the respective risk; details are available in [33]).

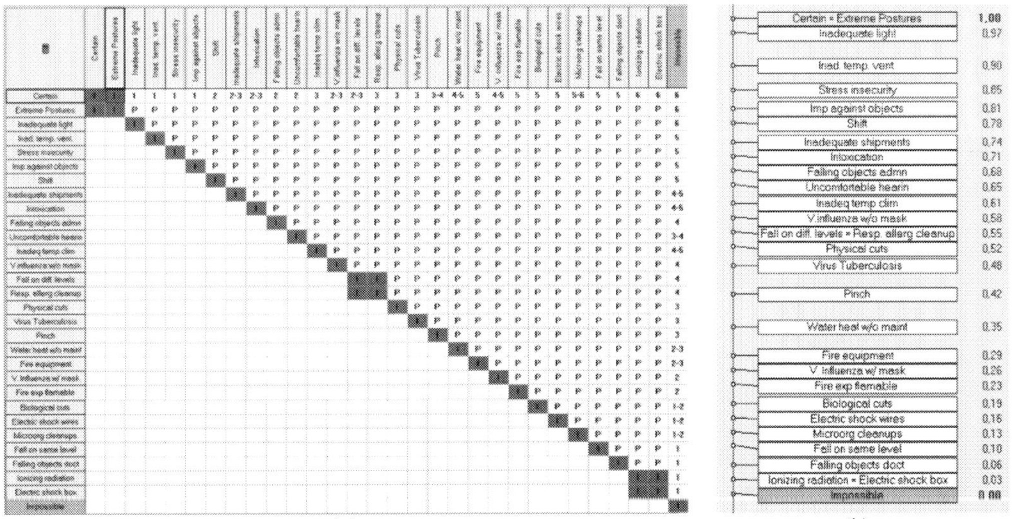

(a) (b)

Figure 5. MACBETH matrix for the representative set of risks (a) and subjective probability scale validated by the DMs (b).

4.4 Risk classification and value risk-matrix

The previous activities enabled the determination of the impact value score and probability of each risk event. Constructing a VRM to be used for evaluating OH&S risks and for analyzing the impact of risk mitigation measures required the identification of risk categories. After discussion with the OHSU group, the five categories depicted in Table 4 (from grade I to grade V) were associated with ranges of combination of the two risk components (impact value score and probability). Then, the OHSU members were asked to classify a wide range of risk event (with different locations in the VRM) in the five categories. With these answers, we were able to adjust iso-curves that separate adjacent categories with MATLAB, by using a cubic spline interpolant. This step also provided a validation of the VRM, as members of the OHSU had to analyse and discuss the combinations of subjective probability, impact value and risk category for a wide range of risk sources.

Based on the impact value scale, the subjective probability scale and on the identification of risk categories, a VRM was designed (see Figure 1). This framework constitutes the basis for the OHSU evaluating OH&S risks and identifying the need for risk mitigation actions.

Table 4. Risk categories defined with the DMs to classify risks.

Categories	Description
Grade I	Recommendation: routine monitoring
Grade II	Recommendation that, as soon as possible, corrective measures are implemented
Grade III	As soon as possible, the corrective actions must be implemented.
Grade IV	Requires immediate corrective actions for its control & dev. of sustainable prevention programs
Grade V	Justifies the immediate closure of the sector, to obtain their elimination or control

4.5 Decision support tool

Once designed the VRM and determined the impact value scores and the probabilities for a representative set of risks, we started constructing a DSS to facilitate the use of the VRM by the OHSU. This DSS, that we designate as "$SAAR_{SST}$" (Portuguese acronym for "*Sistema de Apoio à Avaliação Riscos em Segurança e Saúde no Trabalho*", see Figure 6), works as a Microsoft Excel and Word add-in that contains a personalized code in Visual Basic for Applications. It integrates the functionality of creating a global database of risks and additional information from literature to inform the evaluation of risks. The configuration of the DSS was discussed in detail with the OHSU group so that it contained all the important functionalities and that it was easy to use.

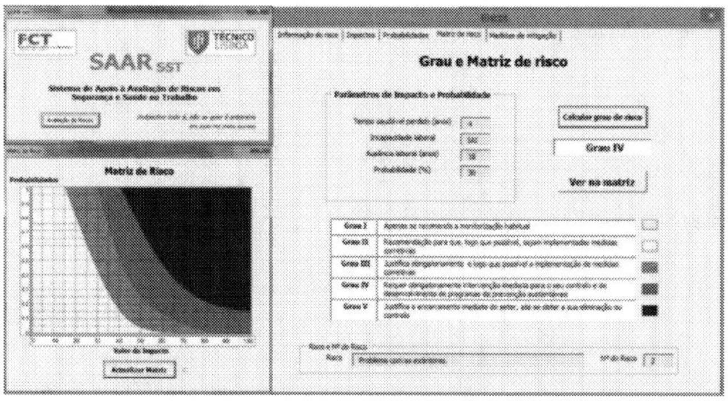

Figure 6. Screenshots of the DSS "$SAAR_{SST}$".

5. Discussion

Departing from an assessment of the problems felt by the OHSU in evaluating OH&S risks and from the RMs methodological problems identified in the literature, this study describes how MCDA concepts and tools were designed and applied – within an interactive learning process – to build a DSS based on a VRM to evaluate OH&S risks.

On the social viewpoint, the OHSU members have found the process for developing the VRM very useful, as it enables and stimulated an identification of problems, a clarification of concepts, and a stimulating learning process. On the technical viewpoint, the VRM corresponds to a flexible and versatile approach that fits in a proper manner the OHSU preferences and overcomes several problems from classical RMs. Therefore, we believe that VRM is a promising methodology that can be applied in many other risk evaluation contexts.

Several aspects are worth of further research. First, VRMs should be developed to other contexts and the VRM for evaluating OH&S risks can be tested with other DMs working in the area. Secondly, further research is needed to prioritize and select mitigation actions in the context of limited resources. We suggest the use of the recent portfolio decision analysis component of the MACBETH DSS, as explained in [9]. Thirdly, it is worth exploring uncertainties associated with the measurement and evaluation of OH&S risks and of mitigation actions.

Acknowledgments

This work is funded by National Funds from the Portuguese Public Budget through FCT – *Fundação para a Ciência e a Tecnologia*, within the project PTDC/EGE-GES/119230/2010. The authors would also like to acknowledge the collaboration of the members of OHSU from the RHA LVT that enabled the development of this study. The authors remain responsible for any omissions and inaccuracies.

References

[1] Alli B 2008 *Fundamental principles of occupational health and safety* (Geneva: International Labour Organization).

[2] Froneberg B 2005 Challenges in occupational safety and health from the global market economy and from demographic change — facts, trends, policy response and actual need for preventive occupational health services in Europe *Scand. J. Work Environ. Health.* **1** 23-27.

[3] Quality and Patient Safety Directorate 2011 Risk assessment tool and guidance (including guidance on application) http://www.hse.ie/eng/.

[4] National Patient Safety Agency 2008 A risk matrix for risk managers http://www.npsa.nhs.uk/nrls/improvingpatientsafety/patient-safety-tools-and-guidance/risk-assessment-guides/risk-matrix-for-risk-managers/.

[5] Administração Regional de Saúde de Lisboa e Vale do Tejo 2010 *Segurança e Saúde no Trabalho: Gestão do risco profissional em estabelecimentos de saúde* Series Segurança e Saúde no Trabalho: Gestão do risco profissional em estabelecimentos de saúde.

[6] Cox Jr LA 2008 What's wrong with risk matrices? *Risk Analysis* **28** 497-512.

[7] Cox Jr LA 2009 *Risk analysis of complex and uncertain systems* (Denver: Springer).

[8] Pickering A and Cowley S 2010 Risk Matrices: implied accuracy and false assumptions *J. Health Saf. Res. Practice.* **2** 9-16.

[9] Oliveira M, Bana e Costa C and Lopes D 2013 Improving risk matrices using multicriteria and portfolio decision analysis Working Paper CEG-IST no. 1/2013 Instituto Superior Técnico (University of Lisbon).

[10] European Commission - IP/08/931 2008 Reducing work-related accidents and illness through better risk assessment http://europa.eu/rapid/press-release_IP-08-931_en.htm#fnB1.

[11] IEC/FDIS 31010. Risk management - Risk assessment techniques. In: International Standard. International Electrotechnical Comission; 2009.

[12] Modarres M 2006 *Risk Analysis in Engineering: Techniques, Tools, and Trends* (Boca Raton: CRC Press).

[13] Bricknell M and Moore G 2007 Health risk management matrix - a medical planning tool *J. R. Army Med. Corps.* **153** 87-90.

[14] Vilaragut J, Duménigo C, Delgado J, Morales J, McDonnell J, Ferro R, Ortiz López P, Ramírez M, Mulas A, Papadopulos S *et al.* 2013 Health physics prevention of accidental exposure in radiotherapy. Risk matrix approach *Health Phys. Soc.* **104** 139-50.

[15] Takeuchi T 2010 Matrix analysis and risk management to avert depression and suicide among workers *BioPsychoSoc. Med.* **4** 1-6.
[16] Grabisch M and Labreuche C 2010 A decade of application of the Choquet integral and Sugeno integrals in multicriteria decision-aid *Annals of Operations Reasearch.* **175** 247-290.
[17] Von Winterfeldt D and Edwards W 1986 *Decision analysis and behavioral research* (Cambridge: University Press).
[18] Phillips L 1984 A theory of requisite decision models *Acta Psychol.* **56** 29-48.
[19] European Agency for Safety and Health at Work 2014 Legislation https://osha.europa.eu/en/topics/riskassessment/legislation.
[20] Vose D 2008 *Risk analysis: a quantitative guide* Wiley).
[21] HCN. Noise and health. In: No. 1994/15E. Health Council of the Netherlands (HCN): Comittee on Noise and Health,: 1994.
[22] Passchier-Vermeer W and Passchier WF 2000 Noise exposure and public health *Environmental Health Perspectives.* **108** 123-131.
[23] Presidência do Conselho de Ministros. Decreto-Lei n.º 503/99 de 20 de Novembro. In: N.º 271. Diário da República - I Série-A; 1999. p. 8241-56.
[24] Stouthard M, Essink-Bot M, Bonsel G, Barendregt J and Kramers P 1997 *Disability weights for diseases in the Netherlands* Series Disability weights for diseases in the Netherlands.
[25] Murray C and Lopez A 1996 *The Global Burden of Disease* (Boston: Harvard University Press).
[26] U.S. Department of Labor 2013 *Nonfatal occupational injuries and illnesses requiring days away from work* Series Nonfatal occupational injuries and illnesses requiring days away from work.
[27] Lopes D, Bana e Costa C, Oliveira M and Morton A 2014 Using MACBETH with the Choquet Integral fundamentals to model interdependencies between elementary concerns in the context of risk management in *Proceedings of the 3rd International Conference on Operations Research and Enterprise Systems, ICORES 2014.* B Vitoriano, E Pinson, F Valente (Ed). (ESEO, Angers, Loire Valley, France: SCITEPRESS) p. 116-26 (digital edition).
[28] Bana e Costa C, Lourenço J, Chagas M and Bana e Costa J 2008 Development of reusable bid evaluation models for the Portuguese Electric Transmission Company *Decision Analysis.* **5** 22-42.
[29] Hubbard D and Evans D 2010 Problems with scoring methods and ordinal scales in risk assessment *IBM J. Res. Dev.* **54** 2:1-2:10.
[30] Wallsten T, Budescu D, Zwick R and Kemp S 1993 Preferences and reasons for communicating probabilistic information in verbal or numerical terms *Bull. Psychon. Rev. Soc.* **31** 135-38
[31] Baron J 2008 *Thinking and Deciding* Cambridge University Press).
[32] Aven T 2008 *Risk analysis: assessing uncertainties beyond expected values and probabilities* Wiley).
[33] Bernardes M 2014 *Building subjective probabilities for health and safety risks with MACBETH* Master Thesis in Biomedical Engineering, Instituto Superior Técnico, University of Lisbon.

Journal of Physics: Conference Series **616** (2015) 012011 doi:10.1088/1742-6596/616/1/012011

A compact formulation for maximizing the expected number of transplants in kidney exchange programs

Filipe Alvelos[1], Xenia Klimentova[2], Abdur Rais[4] and Ana Viana[2,3]

1 Centro Algoritmi / Departamento de Produção e Sistemas, Universidade do Minho, 4710-057 Braga, Portugal

2 INESC TEC, Campus da FEUP, 4200-465 Porto, Portugal

3 ISEP - School of Engineering, Polytechnic of Porto, 4200-072 Porto, Portugal

4 Centro Algoritmi Universidade do Minho, 4710-057 Braga, Portugal

E-mail: `xenia.klimentova@inescporto.pt`

Abstract. Kidney exchange programs (KEPs) allow the exchange of kidneys between incompatible donor-recipient pairs. Optimization approaches can help KEPs in defining which transplants should be made among all incompatible pairs according to some objective. The most common objective is to maximize the number of transplants. In this paper, we propose an integer programming model which addresses the objective of maximizing the *expected* number of transplants, given that there are equal probabilities of failure associated with vertices and arcs. The model is compact, i.e. has a polynomial number of decision variables and constraints, and therefore can be solved directly by a general purpose integer programming solver (e.g. Cplex).

1. Introduction

Kidney transplantation is essential for survival of many patients suffering from chronic kidney disease for which there is no known cure. Kidneys from cadavers used to be the most acceptable source of kidneys for transplantation but it only met a tiny fraction of the demand. Alternative possibilities of transplant existed when there were people willing to donate kidneys to help their loved ones. However, in many cases, kidney recipients and the living donors might be incompatible for success of transplantation because of blood or tissue (physiological) incompatibility.

To bypass this incompatibility deadlock of a recipient-donor pair, the concept of paired kidney exchange between two recipient-donor pairs is utilized increasingly more in practice. In this context, when physiological compatibility is fulfilled, the donor in one pair provides a kidney to the recipient in a second pair in return for the donor in the second pair to provide a kidney to the recipient of the first pair. Two kidney patients may thus benefit from one such "pairwise" exchange of donor kidneys. This concept of kidney exchange can be extended to include more than two pairs and the program facilitating such exchange is generally known as the kidney exchange program.

These programs are typically modelled as a single-objective optimization problem and most past research focused on maximizing the total number of planned transplants [1–3]. Moreover, deterministic mathematical models had been used for optimization which overlooks the likelihood of changes in the pool of recipient-donor pairs (data sets) [4–7]. Changes in the data sets are indeed realistic and should be taken into account as donors can drop out of the pool, recipients may die prior to actual implementation of a scheduled program application, or more rigorous compatibility tests performed after "algorithmic" matching determine that pairs selected for the exchange are incompatible.

In this paper we consider the possibility of such unexpected or last-minute changes in modelling by associating probabilities with the recipient-donor pairs. We describe an integer programming model having polynomial number of variables and constraints what allows being solved directly by a general purpose integer programming solver (e.g. Cplex [8]). For the objective of optimization, we maximize the expected number of successful kidney transplants.

2. Problem statement

Given a directed graph $G(V, A)$, the set of vertices V is the set of incompatible donor-patient pairs. Two vertices i and j are connected by arc $(i, j) \in A$ if a donor from pair i is compatible with a patient of pair j. A cycle in G represents a feasible kidney exchange between two or more pairs. Usually, a limit, denoted by K, is imposed on the length of a cycle to exclude long cycles, which are less desirable due to logistics considerations (since all transplants in a cycle must be done, ideally, at the same time) and also because if a failure occurs, in general, more transplants will be affected in a longer cycle than in a shorter one.

The kidney exchange problem (KEP) for maximizing the number of planned transplants can be defined as follows: *find a packing of vertex-disjoint cycles with length at most K with maximum number of arcs.*

One of the main problems in the implementation of the solution of a KEP instance is its unreliability. Last-minute testing of donor and patient can elicit new incompatibilities that were not detected before, causing some donations in a cycle to be cancelled (so called, positive cross-matching), patients or donors may become unavailable, e.g., due to illness or to backing out, etc. To take into account this fact we consider a problem where a probability of failure is associated to arcs and vertices. We assume that the probability of arc failure is equal, for all arcs. The same holds for vertices:

$$p_i = p^v, \forall i \in V, \quad p_{ij} = p^a, \ \forall (i, j) \in A.$$

It is common practice to allow that vertices in a cycle with failure can be re-matched among them, if a shorter cycle involving only vertices from that cycle can be obtained [9]. Consider the example presented in figure 1 and assume only vertices failure. The presented cycle leads to 3 transplants if none of the three vertices fails, and to 2 transplants (cycle $(1, 2)$) in case of withdrawal of vertex 3 only.

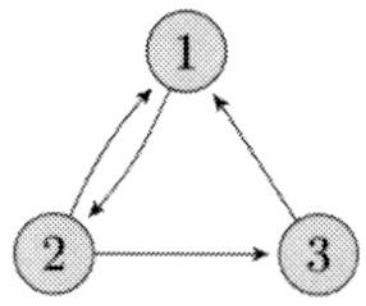

Figure 1. Cycle with 3 vertices.

The expected number of transplants for this cycle is (see [4]):

$$E(c) = 3(1 - p_1)(1 - p_2)(1 - p_3) + 2(1 - p_1)(1 - p_2)p_3$$

and in case of equal probabilities of vertices failure resumes to:

$$E(c) = 3(1 - p^v)^3 + 2p^v(1 - p^v)^2$$

For cycles having similar *configurations*, which may be defined as all *non-isomorphic graphs of given size k having at least a cycle of size k*, the expected value is computed in the same way. The number of different configurations of graphs grows very rapidly with the number of vertices: for $k = 4$ there are 61 different configurations and for $k = 5$ there are 3725 different non-isomorphic graphs.

Denote by T the set of all different configurations of sizes k, with $k = 2, \dots, K$; we index the configurations by t. For the KEP with $K = 4$ there exist 66 configurations of cycles:

- 1 configuration of cycles of length 2: $t = 0$;
- 4 non-isomorphic configurations of cycles of length 3: $t = 1, \dots, 4$;
- 61 non-isomorphic configurations of length 4 with $t = 5, \dots, 65$.

Each configuration $t \in T$ can be represented by arcs of a basic cycle and its set of additional arcs. In figure 2, left, we represent in solid lines a cycle of size 3, composed by arcs (1,2), (2,3), (3, 1). Configurations of this cycle are those involving one, two, or three extra arcs (represented by dashed lines). Similarly, figure 2, right, represents all the possible arcs for a cycle of size 4.

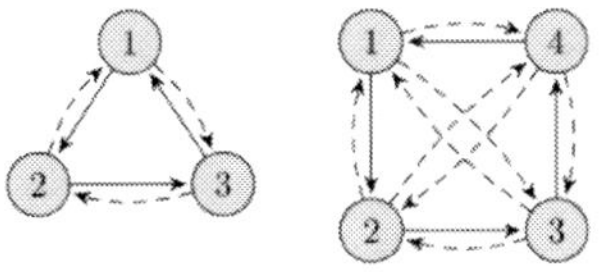

Figure 2. Cycle of length 3 and 4. The basic cycle is presented by solid lines; additional arcs are dashed.

Figure 3 illustrates all configurations of size 3 and three examples of configurations of size 4.

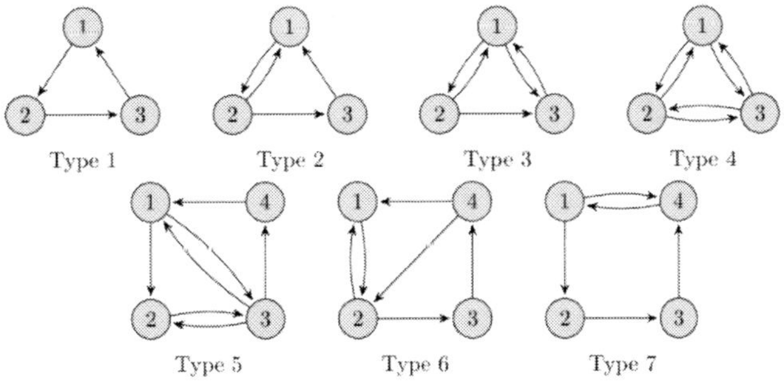

Figure 3. Configurations of 3- and 4-cycles.

Denote by C_t the set of arcs in the basic cycle of configuration t. Denote by Q_t the set of additional arcs, and let R_t be the set of arcs that define a configuration. For example, for configurations of size 3 presented in figure 3: $C_t = \{(1,2),(2,3),(3,1)\}$, $Q_t = \{(2,1),(3,2),(1,3)\}$, $\forall t = 1,\dots,4$, and configurations are given by $R_1 = \{(1,2),(2,3),(3,1)\}$, $R_2 = \{(1,2),(2,3),(3,1),(2,1)\}$, $R_3 = \{(1,2),(2,3),(3,1),(2,1),(1,3)\}$, $R_4 = \{(1,2),(2,3),(3,1),(2,1),(1,3),(3,2)\}$. Similarly, for configurations of size 4 presented in figure 3, we have: $R_5 = \{(1,2),(2,3),(3,4),(4,1),(1,3),(3,1),(3,2)\}$, $R_6 = \{(1,2),(2,3),(3,4),(4,1),(2,1),(4,2)\}$, $R_7 = \{(1,2),(2,3),(3,4),(4,1),(1,4)\}$.

If we have equal probabilities of arc and vertex failures, the expected values for all cycles can be calculated in advance without enumerating them, since they are associated with a tractable number of configurations. The same does not happen if we have different probabilities. In this case, calculating the expected values for all cycles implies enumerating all the cycles, which are in exponential number with respect to the size of the graph, and therefore, they may not be tractable for reasonable sized instances. We say that each configuration defines a type of cycle $t \in T$, and the expected value of that type gives the weight for cycle of a given type — v_t, $t \in T$. Given these weights on cycles the KEP of maximizing the expected number of transplants may be solved by the so-called cycle formulation [3, 10, 11]. However this formulation has exponential number of variables associated to all cycles in the graph of size up to K, that need to be enumerated a priori. On the other hand, the known compact formulations for the KEP [3] are not suitable for solving it, as the presented problem has the weights associated to the types of cycles which may not be transformed into the weights of arcs and vertices. In the next section we present a new compact integer programming formulation that overcomes the problem.

3. Integer programming formulation

In this section we present a new compact, i.e. with a polynomial number of decision variables and constraints, integer programming formulation for maximizing the expected number of transplants in a KEP when equal probabilities of failure are associated to arcs and vertices. Differently from the usual approaches for KEP optimization problems, this new formulation operates with variables on vertices (not arcs) of the graph.

Let N be an upper bound on the number of cycles in any solution (e.g. the number of vertices $|V|$). The cycles in the solution can be represented by an index l, with $l \in L = \{1, \ldots, N\}$. In this case we say that a cycle is *initialized* by vertex l. Given that only cycles with at most K arcs are feasible, preprocessing allows reducing the number of vertices and arcs of the graphs associated with the cycles initialized by each l, which we represent by subgraphs $G^l(V^l, A^l)$ (see [3] for more details). The set of vertices V^l, $l \in L$ is formed by all vertices with an index higher than l, and the vertices that are more distant than the given limit K are eliminated. The set of arcs is given as $A^l = \{(i, j) \in A : i, j \in V^l\}$.

Note that as the non-isomorphic configurations of the graph are considered, the label of each vertex in a configuration can be referred as a position of the vertex of the graph in the configuration. For example, in the graph presented in figure 4 the cycle $(1, 5, 4)$ is a 3-cycle of type 2 (see figure 3) and its vertices have positions 3, 1 and 2, respectively. The cycle $(1, 2, 3)$ is a 3-cycle of type 3 having the vertices in positions 2, 3 and 1, respectively.

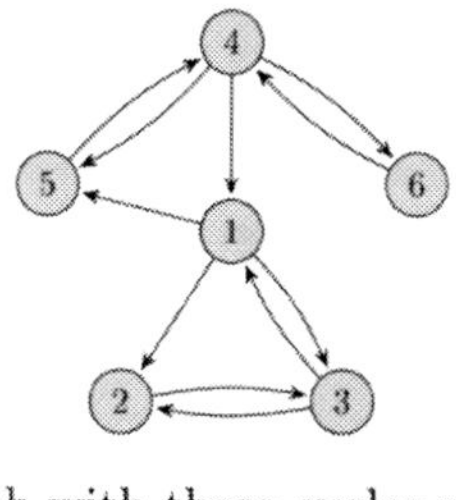

Figure 4. Example of graph with three cycles with different configurations.

The following decision variables are used in the model: $z_{tl} = 1$ if vertex l initialize the cycle of type t, $z_{tl} = 0$ otherwise, where $l \in L$, $t \in T$; $y_i^{qtl} = 1$ if vertex i is selected for position q in the cycle of type t initialized by vertex l, $y_i^{qtl} = 0$, otherwise, $\forall i \in V^l$, $q \in P_t = \{1, \ldots, k\}$, $l \in L$, $t \in T$. Furthermore, we define $\bar{R}_t = (C_t \cup Q_t)/R_t$.

The compact model for the problem under discussion is written as follows:

$$\text{maximize} \sum_{l \in L} \sum_{t \in T} v^t z_{tl} \tag{1}$$

subject to

$$\sum_{t \in T} \sum_{l \in L} \sum_{q \in P_t} y_i^{qtl} \leq 1, \ \forall i \in \bigcup_{l \in L} V^l, \tag{2}$$

$$\sum_{q \in P_t} y_i^{qtl} \leq \sum_{q \in P_t} y_l^{qtl}, \ \forall l \in L, \forall i \in V^l, t \in T, \tag{3}$$

$$z_{tl} = \sum_{q \in P_t} y_l^{qtl}, \ \forall l \in L, t \in T, \tag{4}$$

$$z_{tl} = \sum_{i \in V^l} y_i^{qtl}, \ \forall q \in P_t, l \in L, t \in T, \tag{5}$$

$$y_i^{utl} + y_j^{vtl} \leq 1, \ \forall (i,j) \in \bar{A}^l, \ \forall (u,v) \in R_t, l \in L, t \in T, \tag{6}$$

$$y_i^{utl} + y_j^{vtl} \leq 1, \ \forall (i,j) \in A^l, \forall (u,v) \in \bar{R}_t, l \in L, t \in T, \tag{7}$$

$$z_{tl}, y_i^{qtl} \in \{0,1\}, \ \forall i \in V^l, \ q \in P_t, l \in L, t \in T$$

Constraints (2) guarantee that each vertex can be assigned only to one cycle of one type in one position. Constraints (3) force vertex l to be included into the cycle initialized by l. Constraints (4) put vertex l equal to 1 for some position p if cycle of type t is initialized by l ($z_{tl} = 1$). Constraints (5) can be seen as assignment constraints for the vertex positions. Constraints (6) state that two vertices, i and j with no arc connecting them ($(i,j) \in \bar{A}^l$), cannot be chosen together for positions u and v if an arc between positions u and v is desired ($(u,v) \in R_t$). Constraints (7) state that two nodes, i and j, having an arc connecting them ($(i,j) \in A$), cannot be chosen together for positions u and v if an arc between positions u and v is not desired ($(u,v) \in \bar{R}_t$).

Given that in this problem the *expected* number of transplants is to be maximized, the probability of failure of a cycle is explicitly taken into account in the model.

For the graph presented in figure 4 the optimal solution includes cycles $(1,2,3)$ and $(4,6)$, having $z^{31} = y_1^{231} = y_2^{331} = y_1^{131} = 1$ for 3-cycle and $z^{04} = y_4^{104} = y_4^{204}$ for 2-cycle.

4. Conclusions

The kidney exchange problem is usually treated as a deterministic problem, where all data is considered to be certain, and we want to maximize the number of possible transplants in a pool of incompatible patient-donor pairs. However, in practice, several uncertain events can happen and strongly affect the decisions based on the solutions from deterministic optimization models, e.g. pairs may drop-out of the pool in the interval between algorithm matching and actual transplants, or last minute incompatibility between pairs may be detected.

This paper handles event uncertainty in kidney exchange programs through association of probabilities of arc and vertex failures. The main contribution is the proposal of a compact Integer Programming model, where the maximum expected number of transplants is to be maximized. By assuming that all arcs have the same probability of failure (the same assumption holding for vertices), we can compute *a priori* the expected probability of survival of all types of cycles. This value is in the objective function, weighting cycle selection.

Acknowledgments

This work is financed by the ERDF — European Regional Development Fund through the COMPETE Programme (operational programm for competitiveness), by National Funds through the FCT — Fundação para a Ciência e a Tecnologia (Portuguese Foundation for Science and Technology) within project "KEP - New models for enhancing the kidney transplantation process. FCT ref: PTDC/EGE-GES/110940/2009", by the North Portugal Regional Operational Programme (ON.2 O Novo Norte), under the National Strategic Reference Framework (NSRF), through the European Regional Development Fund (ERDF), and by national funds through FCT within project "NORTE-07-0124-FEDER-000057". This work has also been supported by FCT — Fundação para a Ciência e a Tecnologia (Portuguese Foundation for Science and Technology) within project scope: PEst-OE/EEI/UI0319/2014.

References

[1] de Klerk M, Keizer K, Claas F, Haase-Kromwijk B and Weimar W 2005 *American Journal of Transplantation* **5** 2302–2305

[2] Segev D, Gentry S, Warren D, Reeb B and Montgomery R 2005 *The Journal of the American Medical Association* **293** 1883–1890

[3] Constantino M, Klimentova X, Viana A and Rais A 2013 *European Journal of Operational Research* **231** 57–68

[4] Pedroso J P 2014 *Computational Science and Its Applications ICCSA 2014* (*Lecture Notes in Computer Science* vol 8580) ed Murgante B, Misra S, Rocha A M A, Torre C, Rocha J G, Falcão M I, Taniar D, Apduhan B O and Gervasi O (Springer International Publishing) pp 32–46

[5] Dickerson J, Procaccia A and Sandholm T 2013 *In EC-13: Proc. 14th ACM Conference on Electronic Commerce, June*

[6] Li Y, Kalbfleisch J, Song P, Zhou Y, Leichtman A and Rees M 2011 Optimization and simulation of an evolving kidney paired donation (KPD) program Working Paper Series 90 Department of Biostatistics, University of Michigan http://www.bepress.com/umichbiostat/paper90

[7] Chen Y, Li Y, Kalbfleisch J, Zhou Y, Leichtman A and Song P K 2012 *IEEE Trans. Biomed. Engineering* **59** 1985–1991

[8] IBM 2014 ILOG CPLEX optimizer http://www-01.ibm.com/software/integration/optimization/cplex-optimizer/ (last accessed in September, 2014)

[9] Manlove D and O'Malley G 2012 *Lecture Notes in Computer Science* **7276** 271–282

[10] Roth A, Sönmez T and Ünver M 2007 *The American Economic Review* **97** 828–851

[11] Abraham D, Blum A and Sandholm T 2007 *In: Proceedings of the 8th ACM conference on Electronic commerce, June 13-16* 295–304

A matheuristic for the selection of beam directions and dose distribution in Radiotherapy Planning

T M Obal[1,2], H O Florentino[3], V G Gevert[1], D F Jones[4], D Ouelhadj[4], N M P Volpi[1], and V E Wilhelm[1]
[1]Federal University of Paraná, Curitiba/PR, Brazil
[2]Federal Technological University of Paraná, Guarapuava/PR, Brazil
[3]Biosciences Institute, UNESP, Botucatu/SP, Brazil
[4]CORL, University of Porstmouth, Portsmouth, United Kingdom

E-mail: thalitaobal@utfpr.edu.br

Abstract. In this paper a matheuristic using a combined Genetic Algorithm (GA) and exact method approach is proposed for selecting the position of the beams and dose distribution in Intensity Modulated Radiotherapy Planning (IMRT). GA selects a set of beams, for which the dose distribution is determined in the process of the GA's evaluation, using an optimisation model that is solved by an Interior Point method. Two instances are used to evaluate the performance of the matheuristic, comparing to the optimum solution, in terms of solution and computation time, found using the exact methodology of Branch and Bound. The results show that the matheuristic is appropriate to this problem in the case study proposed, as it is extremely faster than the exact method and also have reached the optimum solution in several experiments done.

1. Introduction

Radiotherapy is one of the most used kinds of treatment to combat cancer. The idea of the treatment is to determine the necessary dose to reduce, or even kill the tumor, whilst making sure not to affect the surrounding healthy organs.

One of the first optimisation models for the radiotherapy problem was proposed by Bahr et al. [3], in 1968. Thereafter, many improvements have taken place in the fields of radiotherapy and Operational Research, to improve the treatment planning.

However, the success of the treatment planning depends on the experience of the physicist in using a trial-error procedure to solve the following problems.

- How to position the gantry and consequently the beam direction (geometric problem);
- How much dosage to deliver through each beam and sub-beam (dose intensity problem);

These are classical problems in radiotherapy planning and the literature reports some papers that use either exact methodologies or metaheuristics to solve them.

One of the most employed exact method to solve these problems is Interior Point. It was used by Obal et al. [19] to solve the multiobjective model of the dose intensity problem. Romeijn et al. [21] proposed a linear approach for the same problem and discussed the hard restrictions of the problem.

Viana et al. [27] used Interior Point to solve the model proposed by [1] - to the dose intensity problem - adding an heterogeneity correction in the composition of different types of irradiated tissues, based on proportions among their different linear attenuation coefficients.

Voet [32] and Breedveld [5] used the algorithm iCycle, that is an algorithm for integrated, multi-criteria optimization of beam angles and profiles, based on *a priori* defined plan criteria, that is called a "wish-list", and is solved using Interior Point.

Other approaches use Branch and Bound or Branch and Cut methodologies [17], [25] and [12]. Souza [25] developed an integer optimisation model to solve the beam directions problem and to evaluate the average dose in the organs at risk. The model was solved using a Branch and Cut algorithm. Lim et al. [17] developed a computational approach to optimise the beam directions and fluence map in IMRT. The authors use an iterative procedure trying to improve the computational performance in comparison with the integer model that was solved with the Branch and Cut methodology.

Gevert et al. [12] proposed a single model for the optimisation of the beam directions and the dose intensity. A set of candidate beams is proposed, and the model inserts an upper bound on the number of beams that could be chosen. The choice is made to optimise the delivery dose in each tissue, finding the distribution of the dose in the beams.

The approach to select a limited number of beam directions is very important, because it determines the gantry position, taking into consideration the limited number of beams imposed by the health plan. That is a clinic reality in the Brazilian health system, although is not in some health institutions in other countries, as discussed in [6].

Moreover, some research has used metaheuristics for radiotherapy problems. Petrovic et al. [20] used a case based reasoning system (CBR), that employs a modified DempsterShafer theory to fuse dose plans suggested by the most similar cases retrieved from the case base, intending to find the best selection of dose distribution and beam directions. After each run of the CBR system, the Simulated Annealing algorithm searches for such a combination of weights for which the difference between the doses suggested by the oncologist/system and doses recommended by the extracted similar cases are minimal.

Simulated Anealling is used by Acosta et al. [1], who propose a model that aims to determine the optimum distribution to each radiation beam, and by Aubry et al. [2].

Another approach was proposed by Hou et al. [15] who developed an optimum solution process in two phases: (1) the optimization of the profiles to a certain set of beams; and (2) the selection of an ideal set of beams based on the result of the objective function to the optimization of the profile. A Dynamic Simulation Algorithm was used to solve the first phase and Genetic Algorithms was used in the second phase.

The research proposed in this paper is a matheuristic to solve a goal programming, which is a hybridization between Genetic Algorithm (GA) and Interior Point. GA is proposed to select the beam directions, and Interior Point to solve the dose distribution problem to GA's selected set of beams.

The experiments have been conducted to improve the computation time that the methodology of Gevert et al. [12]. Although their methodology is robust, it could be very expensive in computation time, which is not desirable due the urgent natural of radiotherapy planning.

2. Problem formulation

Conventionally, the IMRT planning is done on a linear accelerator, with the following components: the couch, the gantry and the collimator (Figure 1).

The linear accelerator is an isocentral machine, that means there is a system formed to the plane of the gantry (that could rotated 360 degrees) and the plane of the couch. The central point of this system is called the mechanical isocentre (Figure 1). This point is first marked on the computational tomography (CT) slices of the patient, and after that, some tags are placed on the body of the patient that serve as reference to position the couch and the gantry. The mechanical isocentre is the base point that represents the centre of the sphere where it is possible to position the beam direction.

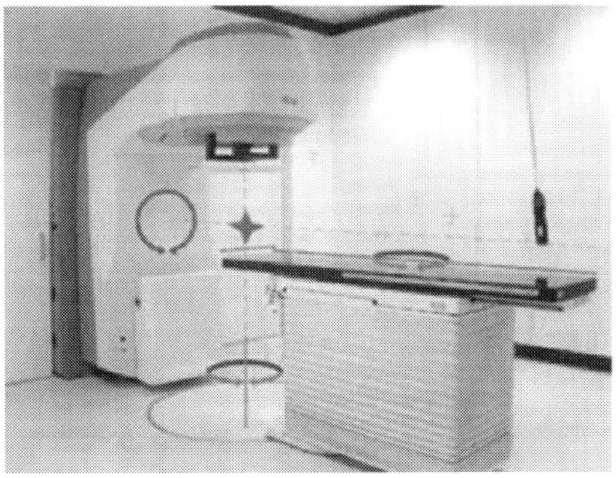

Figure 1. Linear accelerator [18]

The problem is modeled considering a region of the body obtained by the slice of the CT, as shown in Figure 2. The structures selected are: the tumor, in the prostate region; the organs at risk (noble tissue), represented by the head of femur, rectum and the bladder.

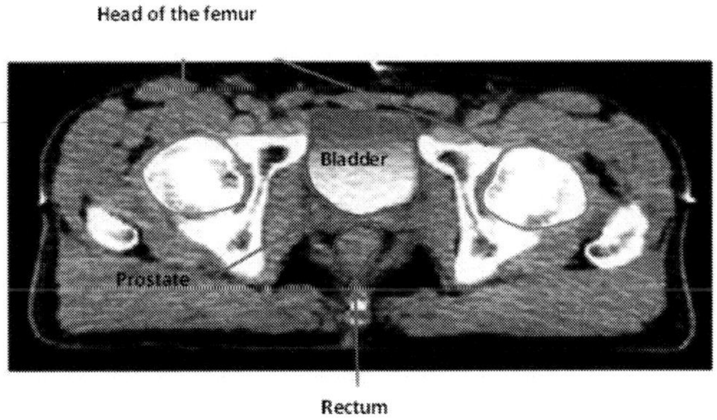

Figure 2. Tomography slice of prostate region

The CT slice is represented by a grid with p pixels, in which each pixel is considered as healthy, noble or tumor tissue.

The central objective of radiotherapy planning is to determine how to design the fluence map so that the dose that reaches the healthy and noble pixels is as small as possible and that the tumor dose is closest to the one prescribed by the Physician.

This is a goal programming problem [16], because it has two clear set of goals: delivery of the necessary dose to the tumor, and least dose possible to the healthy organs. These are mathematical goals and deviation variables should be used, as will be shown in section 3.

The objective of this paper is to develop a matheuristic to find the beam directions and the dose intensity, intending to achieve the treatment goals.

3. Model

Attending to the treatment goals, model (1) is proposed in Obal et al. [19]. This model aims to find the best distribution of the dose for an established set of beams.

$$\begin{cases} min \quad f(\theta^+) = \sum_{i=1}^{pn}(\theta_i^+) & (a) \\ min \quad f(\delta^+) = \sum_{i=1}^{ph}(\delta_i^+) & (b) \\ min \quad f(\epsilon^+) = \sum_{i=1}^{pt}(\epsilon_i^+) & (c) \\ min \quad f(\epsilon^-) = \sum_{i=1}^{pt}(\epsilon_i^-) & (d) \end{cases}$$

subject to

$$\begin{cases} A_nX = S_nB_n + \theta^+ - \theta^- & (e) \\ A_sX = S_sB_s + \delta^+ - \delta^- & (f) \\ A_tX = DB_t + \epsilon^+ - \epsilon^- & (g) \\ (\theta_i^+), (\theta_i^-), (\delta_i^+), (\delta_i^-), (\epsilon_i^+), (\epsilon_i^-) \geq 0 \end{cases} \tag{1}$$

pn, ph and pt are the respective number of noble, healthy and tumor pixels, with $pn+ph+pt = l.c$, and $l \times c$ is the CT slice's size. $X \in \mathbb{R}^{K.m.n}$ represents the emitted dose by each sub-beam, in which $m.n$ is the number of sub-beams in the beam k, with $k = \{1, ..., K\}$.

The dose delivered to the noble and healthy tissue needs to respect the dose limit: $S_n \in \mathbb{R}$ and $S_s \in \mathbb{R}$, respectively. $D \in \mathbb{R}$ is the dose intensity prescribed by the doctor, that needs to be in the tumor structure. $B_n \in \mathbb{R}^{pn}$, $B_s \in \mathbb{R}^{ph}$ and $B_t \in \mathbb{R}^{pt}$, index each kind of tissue: noble, healthy and tumor, respectively.

$A_n \in \mathbb{R}^{pn \times m.n}$, $A_s \in \mathbb{R}^{ph \times m.n}$ and $A_t \in \mathbb{R}^{pt \times m.n}$ represent the absorption dose in each tissue, by the emitted dose of each sub-beam.

The vectors: $\theta^+ \in \mathbb{R}^{pn}, \delta^+ \in \mathbb{R}^{ph}$ and $\epsilon^+ \in \mathbb{R}^{pt}$ represent the deviation of excess dose in the noble, healthy and tumor tissues, respectively; $\epsilon^- \in \mathbb{R}^{pt}$ represent the deviation of deficit dose on the tumor structure; $\theta^- \in \mathbb{R}^{pn}$ and $\delta^- \in \mathbb{R}^{ph}$ are relate to the difference between the dose that arrives in the noble and healthy tissues and its respective bounds.

So, this model has $2.(pn + ph + pt) + K.m.n$ variables and $l.c$ constraints, and aims to find as closer as possible dose D in the tumor structures, which is represented on the functions (c) and (d), and the minimum dose in the noble and health structures, represented in the functions (a) and (b).

The model (1) include the constrains (2) and (3), which aims to determine which beams are optimum to be selected and how much intensity dose in each selected beam is best achieve the optimal radiotherapy plan.

$$\sum_{k=1}^{K} z_k \leq \eta \tag{2}$$

$$0 \leq X_k \leq M.z_k \tag{3}$$

in which K is the possible number of beam direction; η is the maximum number to be select, with $\eta \leq K$; z_k is a binary variable that represent if the k beam is used or not; and $X_k \in \mathbb{R}^{m.n}$ is the sub-vector of X, that represent just the beam k.

The addition of constraints (2) and (3) to the model (1), results in a mixed integer linear programming model. This leads to more computation time of the exact solution process, because of the problem dimension, as will be show in section 5.1. So that, the Genetic Algorithm (GA) is proposed in this research, that make the selection of the beams to be used.

4. Matheuristic for selection of beam directions and dose distribution

This research proposes a hybrid methodology, that uses GA and mathematical programming exact method. For each set of beams that GA proposes, the dose distribution is determine by an Interior Point method solution to model (1).

Based on the principles of natural selection and genetics [11], [4] [14], the Genetic Algorithm is a metaheuristic that generates near-optimal solutions for optimisation problems. It is used due to its ability to find pragmatic solutions in reasonable computation time.

In this algorithm we propose the individual structure as an array of dimension $1 \times K$ where K is the proposed number of possible angular measurements or beam positions. Each column (gene) carries the information 0 or 1 that represent if that beam direction will be used or not, respectively. Figure 3 illustrates the structure of an individual with 8 possible angles at which the first, third, sixth, seventh and eighth beams were selected to make up the treatment plan.

1	0	1	0	0	1	1	1

Figure 3. Example of individual

The initial population is generated using a specific number of individuals. All the individuals created in the initialization are feasible and represent the set of beam directions used in the planning.

Then, this initial population is improved using the following steps [10]:

(i) Fitness evaluation:
The best set of beams is the set which can provide the best possible treatment. The treatment quality is a combination of the set of beams and the dose distribution that it can provide. These problems are dependent, because a better dose distribution depends on the beams selected.
As the beam selection is determined by the GA's individual, the dose distribution problem is solved using model (1) by an Interior Point method, knowing which k-beams will be used. This is the means to evaluate fitness, because better individuals are those which are capable of generating good dose distribution.
The fitness is given by the inverse of the objective function value. A penalty is applied to the infeasible individuals, which hence receive a small fitness.

(ii) Selection: The selection procedure used is roulette-wheel selection ([13], [14]).

(iii) Recombination: One position is randomly chosen and it represents the partition in the two parent individuals that will be divided to create the two new offspring. One of the offspring has the first part of the first parent, and second part of the second parent and the other offspring has the first part of the second parent and the second part of the first parent, as showed in the Figure 4.
It is possible that the created offspring is infeasible, this occurs when all the genes are zero or it has more than desired number of ones. In this case, the infeasible chromosome receives a low value of fitness.
If in this step the created chromosome is equal to another, it suffers a migration process, which means that it was thrown away and another feasible chromosome will take its place.

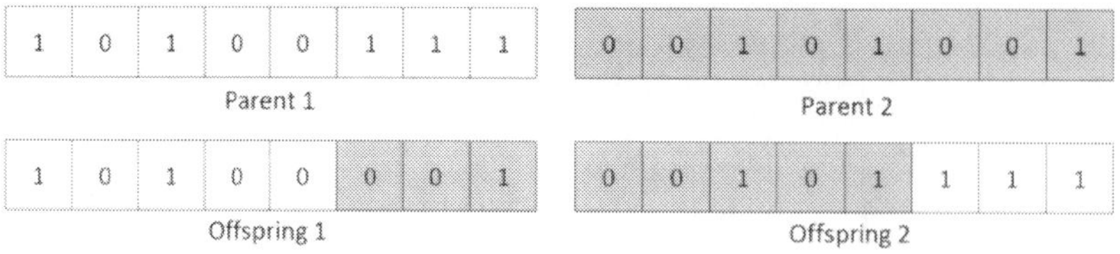

Figure 4. Example of recombination

(iv) Mutation: One gene of individual is chosen and its value is changed, meaning that if it is 1 it becomes 0 and if it is 0 it becomes 1.

Then, the population is replaced and the steps are repeated until a specified number of generations is reached. The experimental results using this methodology will be presented in section 5.2.

5. Experimental results - a case study

The case study used is the cancer of the prostate region, as shown in Figure 2, using one slice of CT. After pre-processing, the image has 11,319 pixels. This image is obtained in Erasto Gaertner Hospital (Curitiba/PR/Brazil), with the project number being 2042.

The experiments are undertaken for two different instances. The first has a set of 8 possible beam directions, with 10 sub-beams each, and up to 6 beam directions could be chosen. The second has a set of 8 possible beam directions, with 10 sub-beams each, and up to 4 beam directions could be chosen. The possible beams directions for the two instances are shown in Figure 5.

In this configuration, the model (1) has 22,718 variables, and 11,319 constraints.

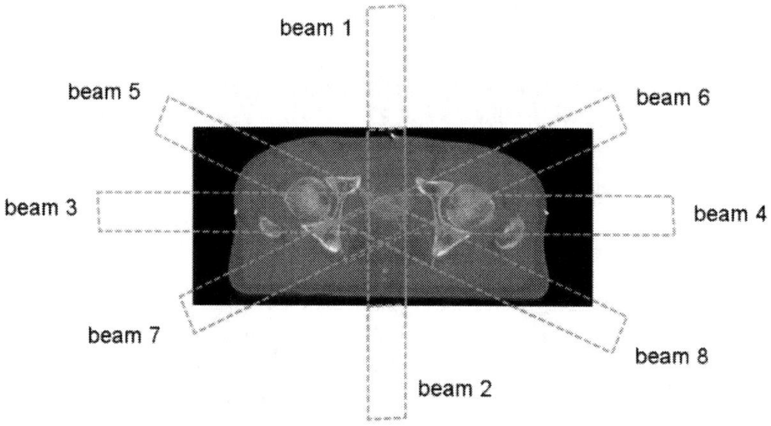

Figure 5. Possible beam directions

Two different solution approaches are used. The first one is the exact methodology, purposed in [12], that will be presented in section 5.1. The exact solution results will be compared to the results obtained by the proposed matheuristic, which will be presented in section 5.2.

The software system Matlab was used, and run on an Intel PC with Core i5 CPU at 1.7 GHz.

5.1. Results of exact methodology

The multiobjective model (1)-(3) is solved with Branch and Bound, using the weighted sum method for the goals, represented on equation (4), with 0.4 as weight to the objective functions (*c*) and (*d*), ie $\gamma_1 = \gamma_2 = 0.4$, and 0.1 as weight to the objective functions (*a*) and (*b*), ie $\alpha = \beta = 0.1$. These weights are chosen based on the experiments done in [12]. Two instances are proposed, using 8 possible beams to select up to 6, for the first instance, and 8 possible beams to select up to 4, for the second instance. The solution of these instances are shown on Table 1.

$$min \quad g = \Big(\gamma_1 f(\epsilon^-) + \gamma_2 f(\epsilon^+) + \alpha f(\theta^+) + \beta f(\delta^+)\Big) \tag{4}$$

Table 1. Results by the exact methodology

Instance	$f(\epsilon^+)$	$f(\epsilon^-)$	$f(\theta^+)$	$f(\delta^+)$	g	Computation time		Beams selected
						seconds	hours	
1	126.52	958.64	731.45	109.47	518.16	94,183	26.16	1, 2, 4, 6, 7, 8
2	63.10	1,152.8	792.61	109.61	576.60	68,658	19.07	1, 2, 5, 7

As presented by Table 1, this methodology is expensive in computational time. The matheuristic proposed which is a hybridisation of GA and Interior Point is used to compare the optimum result obtained with the exact method, and the experimental results are shown in the next section.

5.2. Results of matheuristic

The matheuristic described in section 4 is used for the same instances proposed in section 5.1. For each instance described in Table 2, we run 5 times the experiments, in which the process of selection selects 80% of the individuals, and the mutation probability is 5%.

Table 2. Instances proposed

Instance	Experiment	Number of generation	Population size
1	a	10	5
	b	10	10
	c	5	5
	d	5	10
	e	5	15
2	a	10	5
	b	10	10
	c	5	5
	d	5	10
	e	5	15

In the fitness evaluation, the model (1) is used in the same scenario as in 5.1, with weights: $\gamma_1 = \gamma_2 = 0.4$ and $\alpha = \beta = 0.1$. To solve it, the Interior Point methodology is used, and the results obtained are shown on Table 3.

Table 3 presents the average results for each instance over the five runs, in terms of: objective function for each deviation dose, the weighted sum of the functions (represented by g), the average computation time, the percentage improvement in computation time relative to the exact method and the average number of generations to convergence.

Table 3. Average results of GA

Instance	Experiment	$f(\epsilon^+)$	$f(\epsilon^-)$	$f(\theta^+)$	$f(\delta^+)$	g	Computational time		% improvement in computation time relative to the exact method	Generation of convergence
							seconds	hours		
1	a	103.62	1,035.76	782.90	113.75	545.42	9,536.17	2.65	89.88	*3
	b	118.22	986.74	754.93	105.27	528.00	17,475.13	4.85	81.45	5*1
	c	103.03	1,029.03	790.71	113.44	543.24	4,248.06	1.18	95.49	4*1
	d	116.25	1,005.93	738.00	130.11	535.69	10,240.89	2.84	89.13	3*1
	e	127.23	960.27	738.89	107.41	519.63	18,524.20	5.14	80.34	6*2
2	a	64.04	1,142.47	904.99	117.97	584.90	7,027.90	1.95	89.77	*3
	b	67.25	1,133.00	951.90	115.34	586.82	19,307.37	5.36	71.88	*3
	c	69.57	1,156.31	814.72	233.74	595.20	3,448.84	0.96	94.98	*3
	d	67.03	1,153.32	821.95	134.17	583.75	6,913.31	1.92	89.94	1*1
	e	69.70	1,128.30	914.33	109.09	581.54	9,795.16	2.72	85.74	1*1

*1 Generation of the only convergence case.

*2 Average in 2 times that it converges.

*3 Does not converge.

Table 4. Best results of GA

Instance	Experiment	$f(\epsilon^+)$	$f(\epsilon^-)$	$f(\theta^+)$	$f(\delta^+)$	g	Computational time		% improvement in computation time relative to the exact method
							seconds	hours	
1	a	127.52	960.00	727.54	104.45	518.21	13,887.71	3.86	85.25
	b	126.52	958.64	731.45	109.47	518.16	19,505.65	5.42	79.28
	c	126.52	958.64	731.45	109.47	518.16	2,516.54	0.70	97.32
	d	126.52	958.64	731.45	109.47	518.16	8,646.93	2.40	90.82
	e	126.52	958.64	731.45	109.47	518.16	19,105.17	5.31	79.71
2	a	59.15	1,147.91	839.71	113.91	578.19	5,389.03	1.50	92.15
	b	65.80	1,119.33	965.27	104.14	580.99	16,638.87	4.62	75.77
	c	73.75	1,159.38	828.27	101.12	586.19	3,549.91	0.99	94.83
	d	63.10	1,152.84	792.61	109.61	576.60	7,496.09	2.08	89.08
	e	63.10	1,152.84	792.61	109.61	576.60	10,099.94	2.80	85.29

The results on Table 3 show that the matheuristic reaches the optimum value in 20% of the experiments, for the first instance, and 8%, for the second one.

Table 4 presents the best solution of each experiment in each instance. In the first instance, just one experiment do not reach the optimum, but has a quite small deviation to it (0.01%). In the second instance, for the five experiments done, three do not achieve the optimum, and have, on average, 0.9% deviation.

Both instances present better results using 5 generations and 15 population size, with 0.28% average deviation to the optimum, in the first instance, and 0.85%, in the second. The worst result has average 5.26% deviation to the optimum, in the first instance, and 3.22%, in the second.

The improvement in terms of computational time is very substantial, considering that it is between 80.34% and 95.49%, in the first instance, and between 71.88% and 94.98%, in the second, as presented on Table 3.

Even with a small number of generation, the results are impressive, whereas the experiments that converge have done it in less than 6 generations.

The population size is important to the convergence process, because better results are found with its increase, but have the disadvantage of also increase the computational time.

The presented instances shows the quality of the matheuristic to solve the combined beams selection and dose distribution problem in radiotherapy planning. It is faster than the exact method and finds the global optimum or very close to it in many executions done. More experiments in terms of crossover and mutation process could be conducted in order to improve the GA in this case study. Accordingly, other instances could be tried.

6. Conclusion

The planning of radiotherapy is a process that could be improved with operational research techniques. This paper introduces a methodology to find the beam directions for a set of possible beams to be chosen, and the optimum distribution dose for the selected set of beams.

We have conducted experiments to evaluate the performance of the proposed matheuristic in terms of the solution quality and computation time, comparing it to the exact method. The matheuristic results obtained a great improvement in computation time, and good quality solutions.

Although the matheuristic has achieved a substantial improvement in computation time, we still believe that it could be further improved by using a single population metaheuristics. So, a future work is to investigate other single solution metaheuristics to solve the problems of beam directions and dose distribution, with the intent to improve the computation time of the matheuristic proposed in this paper.

Inclusion of non-coplanar beams with CT 3D configuration is also crucial to the possibility of the use of these methodologies in radiotherapy planning in hospitals.

Acknowledgments

We acknowledge the collaboration of the staff of the radiotherapy Erasto Gaertner Hospital, Curitiba/Paraná/Brazil, having project number 2042 approved by the Committee of Ethics in Research.

Thalita Monteiro Obal gratefully acknowledges financial support from CAPES-Brazil.

References

[1] Acosta R, Brick W, Hanna A, Holder A, Lara D, Mcquilen G, Nevin D, Uhlig P and Salter B. 2009 Radiotherapy optimAl Design: An Academic Radiotherapy Treatment Design System. *Operations Research/Computer Science Interfaces* **47** 401-425

[2] Aubry J, Beaulieu F, Sevigny C, Beaulieu L and Tremblay D. 2006 Multiobjective optimization with a modified simulated annealing algorithm for external beam radiotherapy treatment planning. *Medical Physics* **33** 4718-29.

[3] Bahr G K, Kereiakes J G, Horwitz H, Finney R, Galvin J and Goode, K. 1968 The method of linear programming applied to radiation treatment planning. *Radiology* **91** 686-693.

[4] Bremermann H J. 1958 The evolution of intelligence. The nervous system as a model of its environment. Technical Report No. 1, Department of Mathematics, University of Washington, Seattle, WA.

[5] Breedveld S. 2013. *Towards Automated Treatment Planning in Radiotherapy: A Mathematical Approach to Automated and Integrated Multi-Criterial Optimization of Beam Angles and IMRT Fluence Profiles.* Erasmus University Rotterdam, The Netherlands.

[6] Breedveld S, Storchi P R, Voet P W J and Heijmen B J M. 2012 iCycle: integreted, multi-criterial beam angle and profile optimization for generation of coplanar and non-coplanar IMRT plans *Medical Physics* **39** (2) 951-963

[7] Breedveld S, Storchi P R and Heijmen B J M. 2009 The equivalence of multi-criteria methodos for radiotherapy plan optimization *Physics in Medicine and Biology* **54** 7199-7209

[8] Breedveld S, Storchi P R, Keijzer M and Heijmen B J M. 2007 Automated adjustment of voxel-dependent importance factors in inverse planning *Proceedings of the XVth International Conference for the Use of Computers in Radiation Therapy (Toronto, Canada)* **1** 232-236

[9] Breedveld S, Storchi P R, Keijzer M and Heijmen B J M. 2006 Fast, multiple optimizations of quadratic dose objective functions in IMRT *Physics in Medicine and Biology* **51** 3569-3579

[10] Burke E K and Kendall G. 2005 *Search Methodologies: introductory Tutorials in Optimization and Decision Support Techniques.* Springer, United States.

[11] Fraser A S. 1957 Simulation of genetic systems by automatic digital computers. II: Effects of linkage on rates under selection, Austral. J. Biol. Sci. 10:492:499.

[12] Gevert V G, Obal T M, Volpi N M P and Wilhelm V E. 2013 Um modelo de otimização integrado para os problemas da escolha de feixes de radiação e do mapa de fluência em radioterapia. Simpósio Brasileiro de Pesquisa Operacional, XLV, 2013. Natal/RN. *Anais.* Sociedade Brasileira de Pesquisa Operacional.
[13] Goldberg D E. 1989. Genetic Algorithms in Search Optimization and Machine Learning. *Addison-Wesley*, Reading, MA.
[14] Holand J H. 1975. Adaption in Natural and Artificial Systems. University of Michigan Press, Ann Arbor, MI.
[15] Hou Q, Wang J, Chen Y and Galvin, M. 2003 Beam orientation optimization for IMRT by a hybrid method of the genetic algorithm and the simulated dynamics. *Med. Phys.* **30** 23607.
[16] Jones D. and Tamiz, M. 2010 *Pratical Goal Programming.* International Series in Operations Research & Management Science **141** Springer.
[17] Lim G, Choin J and Mohan R. 2008 Iterative solution methods for beam angle and fluence map optimization in intensity modulated radiation therapy planning. *OR Spectrum* **30** 289-309.
[18] Lopes M C. 2007 Um século de terapia com radiaçáo *Gazeta de Física* **30** 3 Coimbra
[19] Obal T M, Volpi N P and Miloca S A. 2013 Multiobjective approach in plans for treatment of cancer by radiotherapy. *Pesquisa Operacional* **33** (2) 269-282.
[20] Petrovic S, Mishra N and Sundar S. 2011 A novel case based reasoning approach to radiotherapy planning. *Expert Systems with Applications* **38** 1075910769.
[21] Romeijn H E, Ahuja R K, Dempsey J F and Kumar A. 2006 A New Linear Programming Approach to Radiation Therapy Treatment Planning Problems. *Operations Research*, **54** (2) 201-216.
[22] Rossi L, Breedveld S, Heijmen B, Voet P, Lanconbelli N and Aluwini S. 2012 On the beam direction search space in computerized non-coplanar beam angle optimization for IMRT - prostate *Physics in Medicine and Biology* **57** 5441-5458
[23] Shao L. 2008. Multiple Objective Linear Programming in Radiotherapy Treatment Planning Thesis of doctory, Department of Engineering Science, School of Engineering, University of Auckland.
[24] Song X, Petrovic S and Sundar S. 2007 A Case-Based Reasoning Approach to Dose Planning in Radiotherapy, Workshop Proceedings, *The Seventh International Conference on Case-Based Reasoning ICCBR 07*, Belfast, Northern Ireland, 348-357.
[25] Souza W D, Meyer R R and Shi L. 2004 Selection of beam orientations in intensity-modulated radiation therapy using single-beam indices and integer programming. *Phys Med Biol.* **49** (15) 3465-81.
[26] Talbi E G. 2009 Metaheuristics: from desing to implementation. *John Wley & Sons*, New Jersey.
[27] Viana R S, Florentino H O, Lima E A B F, Fonseca P R and Homem T P D. 2011 Heterogeneity correction in the construction of optimized planning in radiotherapy using linear programming *Pesquisa Operacional* **31** (3) 565-578.
[28] Voet P W J, Breedveld S, Dirkx M L P, Levendag P C and Heijmen B J M. 2012 Integrated multicriterial optimization of beam angles and intensity profiles for coplanar and noncoplanar head-and-neck IMRT and implications for VMAT *Medical Physics* **39** (8) 4858-65 .
[29] Voet P W J, Dirkx M L P, Breedveld S, Fransen D, Levendag P C and Heijmen B J M. 2013 Toward fully automated multicriterial plan generation: a prospective clinical study *International Journal of Radiation Oncology Biology Physics* **85** (3): 866-72.
[30] Voet P W J, Dirkx M L P, Breedveld S and Heijmen B J M. 2013 Automated generation of IMRT treatment plans for prostate cancer patients with metal hip protheses: comparison of different strategies *Medical Physics* **40** (7) 071704.
[31] Voet P W J, Dirkx M L P, Breedveld S, AL-Mamgani A, Incrocci L and Heijmen B J M. 2014 Fully automated volumetric modulated arc therapy plan generation for prostate cancer patients. *International Journal of Radiation Oncology Biology Physics* **88** (5) 1175-79.
[32] Voet P W J. 2014. *Automation of contouring and planning in radiotherapy.* Erasmus University Rotterdam, The Netherlands.

Journal of Physics: Conference Series **616** (2015) 012013 doi:10.1088/1742-6596/616/1/012013

Retrieval with Clustering in a Case-Based Reasoning System for Radiotherapy Treatment Planning

Gulmira Khussainova, Sanja Petrovic, Rupa Jagannathan

Operations Management and Information System Division, Nottingham University Business School, University of Nottingham, Nottingham NG8 1BB, UK

E-mail: psxgk@nottingham.ac.uk, sanja.petrovic@nottingham.ac.uk, rupa.jagannathan@nottingham.ac.uk

Abstract. Radiotherapy treatment planning aims to deliver a sufficient radiation dose to cancerous tumour cells while sparing healthy organs in the tumour surrounding area. This is a trial and error process highly dependent on the medical staff's experience and knowledge. Case-Based Reasoning (CBR) is an artificial intelligence tool that uses past experiences to solve new problems. A CBR system has been developed to facilitate radiotherapy treatment planning for brain cancer. Given a new patient case the existing CBR system retrieves a similar case from an archive of successfully treated patient cases with the suggested treatment plan. The next step requires adaptation of the retrieved treatment plan to meet the specific demands of the new case. The CBR system was tested by medical physicists for the new patient cases. It was discovered that some of the retrieved cases were not suitable and could not be adapted for the new cases. This motivated us to revise the retrieval mechanism of the existing CBR system by adding a clustering stage that clusters cases based on their tumour positions. A number of well-known clustering methods were investigated and employed in the retrieval mechanism. Results using real world brain cancer patient cases have shown that the success rate of the new CBR retrieval is higher than that of the original system.

1. Introduction

Every day in England 11 people are diagnosed with brain cancer and it claims 9 lives [1]. The process from diagnosis to treatment contains many steps. Medical images such as computed tomography (CT) and magnetic resonance imaging (MRI) are taken to detect the location of the tumour. After the tumour and the organs surrounding the tumour, so called organs-at-risk (OAR) are outlined the treatment planning process starts. In radiotherapy treatment planning the aim is to deliver sufficient radiation to kill tumour cells while trying to spare healthy tissues in the tumour surrounding area.

Radiotherapy treatment planning is usually a trial and error process and needs the expertise of one or more experienced medical physicists and oncologists. It can take up to 3-4 days depending on the complexity of the plan. Developing a treatment plan depends highly on the expertise and experience of the medical staff. Most software applications that are in use to facilitate radiotherapy treatment planning are based on optimisation methods [2, 3, 4, 5]. However the drawbacks of such methods is that they do not utilise the experience of the medical staff, which can provide valuable information when designing a treatment plan for an individual patient.

Case-based reasoning (CBR) uses previous experience to understand and solve new problems based on similar problems solved in the past [6]. When creating a treatment plan for a new patient medical physicists try to retrieve/remember similar cases that they have encountered in the past along with their treatment plans and try to map it to the new case. Therefore, using CBR seems to be a natural decision in developing an application that will help medical staff to design a treatment plan. The purpose of an automated software application system is to save medical physicists' time. In addition, such system could be used to train junior medical staff.

The research presented in this paper builds on the CBR system for 3D conformal radiotherapy treatment planning of brain cancer which was developed in collaboration with the Nottingham University Hospitals Trust, NHS, Nottingham City Hospital Campus and is described in Jagannathan et.al. [7]. Each case in the case base consists of the case attributes that describe the relationship between the tumour and OARs of that patient and the treatment plan that was used to treat the patient. The treatment plan consists of the number of radiation beams and beam angles. Given a new patient case the CBR system computes the similarity between the attributes of the new case and the cases in the case base and returns the most similar case along with its treatment plan. This treatment plan can then be adapted by medical physicists to meet the specific requirements of the new patient. Adaptation of the retrieved solution is a crucial part of the CBR process that ensures that the solution can be applied to the new case.

The CBR system was tested by medical physicists on new patient cases. However, the results showed that the treatment plans of some of the retrieved cases could not be adapted to be applied to the new case as their tumour position was very different to the tumour position of the new case resulting in highly unsuitable beam angles.

Treatment planning aims to deliver the maximum of the prescribed radiation on the centre of the tumour. Therefore, beams placement is largely determined based on the position of the tumour and points where beams intersect should be placed at the centre of the tumour.

In this paper, we revised the existing CBR retrieval mechanism to retrieve cases with more suitable tumour positions. It is not expected to find a case in the case base with the tumour positioned in the exact same location as the new case. The aim is to consider only the cases whose tumour positions is very close to the new case so that beam angles in the retrieved treatment plan can be adapted. For this purpose cases are clustered by their tumour positions so that during retrieval the system considers only the cases that are in the same cluster as the new case. The results of the revised CBR system are compared with the previously developed CBR system.

The continuation of this paper is as follows. Section 2 contains a description of the developed CBR system and the insights learned from the CBR validation. Section 3 describes methods implemented for determining the optimal number of clusters for tumour position and clustering itself. Section 4 presents the computational results of clustering of cases in the current case base. Section 5 compares the performance of the original and revised CBR systems. The paper finishes with the conclusion in section 6.

2. Background and related work

This section will introduce the existing CBR retrieval system, the insights learned from its use by medical physicists and the required revisions to the existing retrieval mechanism to improve the suitability of the retrieved treatment plan to the new case.

2.1. *Existing CBR system*

The CBR system has been implemented to facilitate in developing a radiotherapy treatment plan for brain cancer patients. It contains a case base of 86 brain cancer patient records received from Nottingham City Hospital containing descriptions of patient cases and their treatment plans.

In the Nottingham City Hospital, after examining the image files of patients the planning treatment volume (PTV) containing the tumour and organs at risk (OAR) are outlined in the treatment planning system by the oncologists. The following OAR are considered in the treatment plan: the right and left lens of the eyes, right and left optical nerves, the spinal cord, optic chiasm and brainstem. The treatment plan is generated by medical physicists in a trial-and-error fashion by placing beams on the patient image in the treatment planning system to achieve an acceptable dose of irradiation.

In the existing CBR system the case most similar to a new case is retrieved using a similarity measure based on the spatial relationship between the tumour and OAR, the volume of the tumour and the ratio of the tumour to the patient body. Not all of the OAR mentioned above are present in all cases. Depending on the tumour location some OAR are absent as they are not affected by radiation beams. For example, if the tumour is located in the front part of the head then the brainstem might not be an OAR as no risk is posed to it due to its large distance to the tumour. First, filtering of the case base is performed, which considers for retrieval only cases with the same OARs as the new case. Then the similarity between the new case and the filtered cases in the case base is computed using a weighted nearest neighbour (wNN) similarity measure with local weights [8]. The retrieval consists of two phases. First, the retrieved case suggests the number of beams, and then among the remaining filtered cases with this number of beams the most similar one suggests the beam angles.

In consultation with medical physicists the following case attributes that affect the treatment plan configuration were identified:

V -volume of the tumour (given in mm).

A - angle between the line connecting the origin of the coordinate system with the centre of the tumour and with the centre of OAR (0-360).

E -minimum distance between the edge of the tumour outline and OAR outline (mm).

R - the ratio of the tumour volume to the volume of patient's body (Body - Tumour volume ratio).

Dt - minimum distance between the body edge and the tumour outline (Body - Tumour volume distance) (mm).

P - the relative position of the tumour to OAR (Tumour - OAR spatial relationship). It contains six labels: left, right, posterior, anterior, superior and inferior.

The treatment plan of the retrieved case is applied to the new patient. At the current stage of development, the CBR system includes two treatment plan parameters: the number of radiation beams which are applied to the tumour and the angles of those beams. Beam angles are defined by the gantry angle, patient couch angle and collimator angles.

The treatment plans in the case base were generated by medical physicists and validated by the oncologists. Our aim is to mimic the way medical physicists generate treatment plans, and at this stage of the project we use the plans generated by them to evaluate our CBR system. We employed the leave-one-out cross validation technique to determine the accuracy of the retrieval system and the similarity measure by repeatedly removing a case with a known solution from the case base, so called target case, retrieving the most similar case from the remaining cases in the case base and comparing the solution of the retrieved case with the actual solution of the target case. The difference between two solution forms the retrieval error. The average error obtained by making each case in this fashion consecutively the target case gives an indication of the performance of the similarity measure. This method is known as the leave-one-out cross validation technique. The error is calculated in terms of the difference between the beam angle(BA) error and the beam number(BN) error in the treatment plans of the retrieved and target cases. The BA error is calculated based on the difference between the beam angles and the BN error, by the difference in beam numbers between the treatment plans of the retrieved and the target cases. The aim is to retrieve a treatment plan with BN error=0 and minimal BA error.

Journal of Physics: Conference Series **616** (2015) 012013 doi:10.1088/1742-6596/616/1/012013

2.2. CBR system validation

The developed CBR system was presented to medical physicists who used it for the treatment planning of 6 new patients. This section describes the insights learned from this validation.

In all 6 cases, the system retrieved treatment plans with the same beam number as the new case. For 2 out of the 6 new patient cases the retrieved cases provided treatment plans that were a good starting point for adaptation; for 2 cases the retrieved cases suggested treatment plans with some acceptable beam angles and some beam angles that medical physicists would not use; finally, for the last 2 new cases the suggested beam angles could neither be used nor adapted for the new case.

When the beams are placed the radiation is focussed at the point, known as the isocentre, where all beams meet and ideally this point should be placed at the centre of the tumour. In all cases in our case base the tumour position in the patient case and the isocentre positions in the treatment plan coincide.

When adapting cases medical physicists noted that the isocentre of the retrieved treatment plan was often misplaced with respect to the tumour position centre of the new patient. The tumour position relative to OAR is captured in the similarity measure by attribute P described earlier. However, this attribute was not sufficient to define the tumour position as the isocentre of the suggested beam in the retrieved treatment plan did not match the tumour centre of the new case.

The tumour position is important as it defines how and from which side of the head beams will enter, e.g. if the tumour is located in the left side of the head then the beams are placed to enter the head from the left side and having beams enter from the right side would mean passing through a lot of healthy tissues. In the cases where the retrieved beams are misplaced in such a manner it is easier to design a new treatment plan from scratch than to adapt the retrieved treatment plan. Therefore, in order to avoid retrieving such non-adaptable treatment plans, it was decided to take into account the tumour position of cases during the CBR retrieval system.

The evaluation of the retrieved treatment plans for the six new cases showed that that metrics used to compute the retrieval error in terms of BA and BN do not give an indication of how suitable the retrieved plan is for adaptation. It is possible that the retrieved treatment plan is difficult to adapt or unsuitable for adaptation in spite of having a small retrieval error. For this reason, the authors developed an adaptation-guided retrieval mechanism, where the modified CBR system retrieves a case that is not only the most similar to the new case but also can be adapted to meet the requirements of the new patient. The results of this work are in the process of publication.

2.3. Tumour position and isocentre for the retrieval mechanism

The aim is to match the isocentre of the retrieved treatment plan with the tumour position of the target case. If the tumour positions of the retrieved and target cases match then the isocentre of the retrieved treatment plan should be a good match for the target case. After the OAR filtering, we filter cases based on their tumour positions. However, tumour positions usually do not match exactly in different cases. Thus, we look for a case to retrieve with a tumour position located close to the tumour position of the new case. All cases in the case base are clustered based on their tumour position. Given a target case, the system finds all cases in the case base with the same tumour position cluster as the target case. Only these cases are available for the retrieval. The system then computes the similarity between the target case and these cases, and retrieves the most similar case.

3. Clustering methods

We use the k-means clustering algorithm [9] to find the clusters of cases based on their tumour positions. K-means is a classical robust clustering technique used when a number of clusters in the dataset is fixed a priori [10]. The k-means clustering algorithm divides M data samples in N dimension into k clusters so that the average squared distance from data points to the cluster centre, referred to as mean, is minimal. The k-means algorithm is iterative and in every iteration each data point is assigned to the closest cluster mean and then the means are recalculated. The optimal solution is found when the assignments no longer change the distance between data points and the cluster mean.

In the anatomical coordinate system that is used in radiotherapy treatment planning the X coordinate increases from the left side of the body towards right, and thus, it defines left and right position of the tumour, the Y coordinate increases towards the anterior part of the body and defines anterior and posterior position of the tumour and the Z coordinate increases towards the superior part of the body and therefore determines superior and inferior position of the tumour. The tumour position in the head/brain described by 3 coordinates: X, Y and Z is shown in Figure 1.

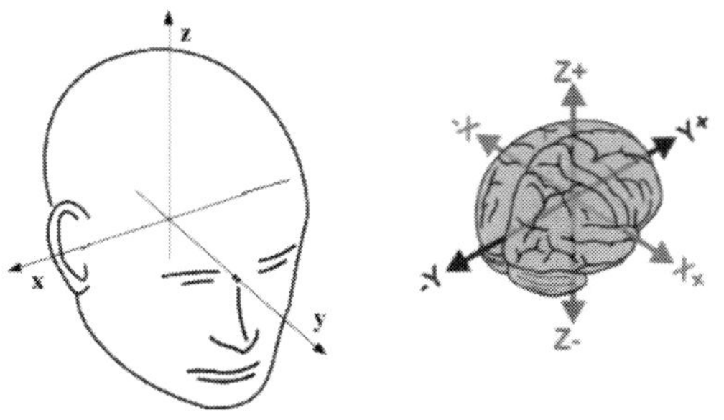

Figure 1. The head and brain anatomical coordinate system [11, 12].

We need to find the number of clusters to use with our data set first. Based on the coordinate system described in Figure 1 we can identify 8 regions in the brain: (Left, posterior, superior), (Left, posterior, inferior), (Right, posterior, inferior), (Right, posterior, superior), (Left, anterior, inferior), (Right, anterior, superior), (Left, anterior, superior), (Right, anterior, inferior).

Based on this we can cluster our data set into 8 clusters. However, our data set might not be distributed evenly and in some clusters there might not be any cases or cases can fall into just small number of clusters.

Therefore, it is required to find the optimal number of clusters based on our data set. In order to define the number of clusters we make use of off-the-shelf techniques available in *R* packages. *R* is an open source computing environment for statistical analysis of a data set.

There are many methods to suggest the optimal number of clusters in the data set. Some of them are designed for large data sets and are therefore, not suitable for our case base. To find the optimal number of clusters we applied 3 clustering methods to the data set. In particular, first we used a heuristic clustering method, known as partitioning around medoids (PAM) [13]. They partition data into clusters in such a way as to minimise the distance between points labelled to be in a cluster and a point designated as the center of that cluster. Heuristic methods look for a partitioning that is good enough for a given data set. Second, we used model-based clustering [14] method which belongs to the class of probability based methods. Probability models analyse the data set and make assumptions about clusters. Both of these 2 methods described look for normally distributed clusters. Thus, finally, we used Calinski-Harabasz clustering [15], which does not look for normally distributed clusters compared to the previous methods.

4. Clustering computational results

This section compares the results of methods for determining optimal number of case clusters.

4.1. PAM clustering results

The PAM method suggested 2 clusters based on tumour position data illustrated in Figure 2. Component 1 and 2 are principal components which define the distribution of the data. The number of components is equal to the number of found correlated data points.

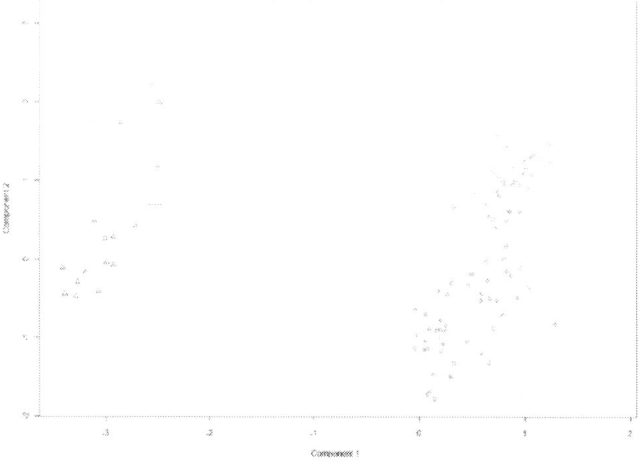

Figure 2. Cases clustered with PAM method

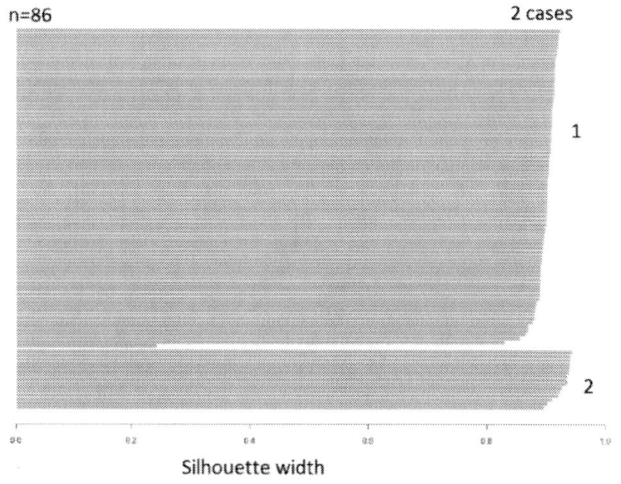

Figure 3. PAM silhouette obtained with cases from the case base

PAM provides a novel graphical display, the silhouette plot, which allows the user to select the optimal number of clusters. Silhouette plot in Figure 3 shows the clustering measure for each case and how it fits into the assigned cluster. This is done by comparing how close the case to other cases in its own cluster and cases in other clusters. Each line on the silhouette plot corresponds to a case. The Y axis represent cases and X axis shows how well the cases fit the assigned cluster. For example, values near 1 mean that the corresponding case fits well into the cluster and values near 0 mean that the case could belong to another cluster. In each cluster the value is displayed from smallest to largest. The information in the silhouette plot is used to determine the optimal number of clusters. For the current case base the silhouette plot suggests 2 clusters. We can see that most of the cases belong to a cluster 1, and a smaller number of cases belong to cluster 2. The average silhouette width of 0.9 suggests that 90 % of cases clustered were fit well by the suggested method.

4.2. Model-based clustering results
Model-based clustering with Bayesian probability clustered cases into 10 different probability models.

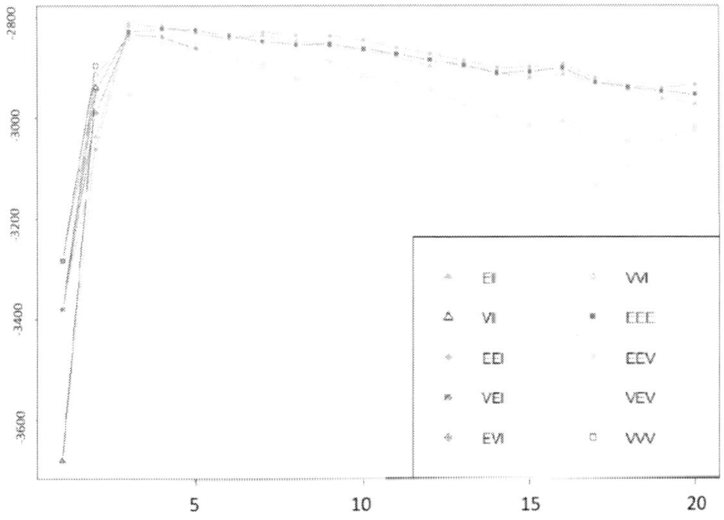

Figure 4. Result of Model Based Clustering

Figure 4 shows that all models converge to a point which suggests 3 to be the optimal number of clusters. The X axis represent the number of clusters and the Y axis their distribution in the space based on Bayesian probability.

4.3. Calinski-Harabasz clustering results
The Calinski-Harabasz clustering analysis presented in Figure 5 suggests 10 clusters to be the optimal solution. In Figure 5 the X axis represents the number of cases and the Y axis represents the number of clusters. Different shades represent different case clusters. The tumour position data of all cases in the case base grouped in 10 clusters can be seen from Figure 6.

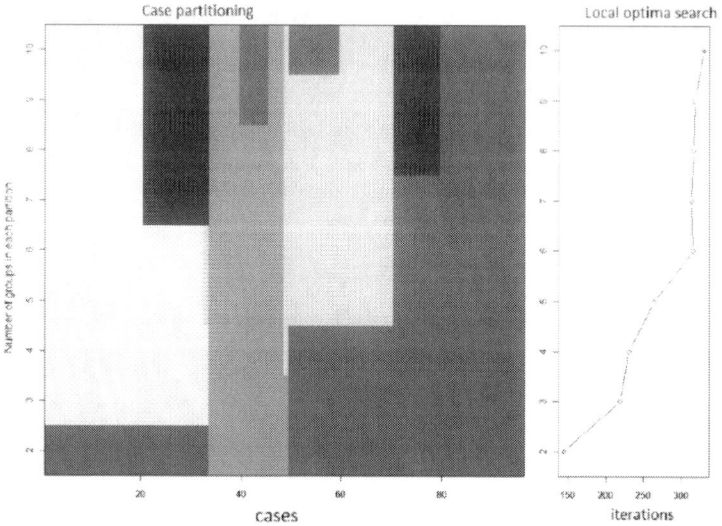

Figure 5. Calinski-Harabasz clustering analysis

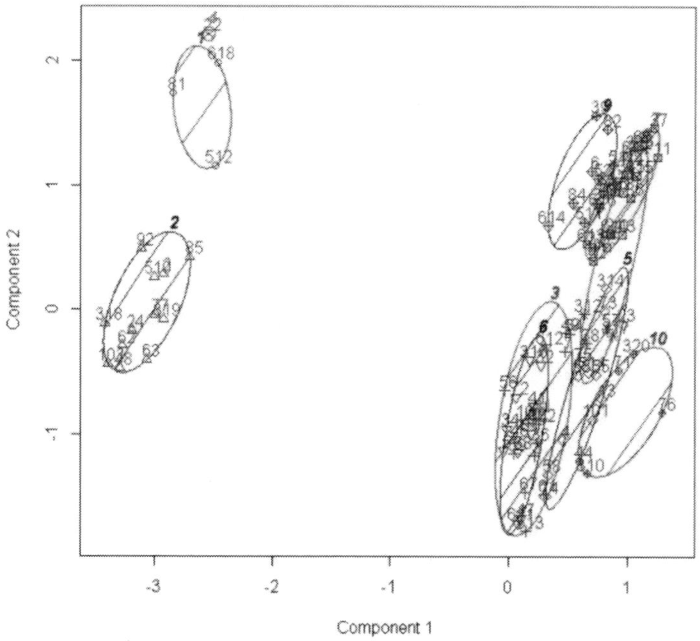

Figure 6. Cases clustered into 10 clusters as suggested by Calinski-Harabasz

The suggested number of clusters from the three methods above, 2, 3, and 10, were used by k-means algorithm to cluster cases by their tumour positions. We also consider clustering into

8 predefined clusters based on the geometry of a brain anatomical coordinate system given in Figure 1.

5. Revised CBR retrieval results and discussion

Experiments were carried out to assess the clustering of cases based on their tumour positions and number of clusters on the retrieval mechanism performance. Experiments were run using 2, 3, 8 and 10 clusters. Leave-one-out cross validation method was used to measure the success rate of the CBR system.

The Euclidean distance between two isocentre points (X_1, Y_1, Z_1) and (X_2, Y_2, Z_2) is calculated using the formula below:

$$d(isocentre_1, isocentre_2) = \sqrt{(X_2 - X_1)^2 + (Y_2 - Y_1)^2 + (Z_2 - Z_1)^2} \quad (1)$$

Table 1 presents the comparison between the original and revised CBR systems. In addition to BN and BA errors we also measure the distance between the isocentre of the retrieved treatment plan and known original treatment plan of the target case.

Table 1. Comparison between retrieval results of the original CBR system and the revised CBR system with clustering

CBR system	Average isocentre distance(mm)	Average BN error	Average BA error	Success rate change w.r.t BN%	Success rate change w.r.t BA%
Original CBR	207.87	0.4	24.36	-	-
CBR with 2 clusters	107.5	0.4	24.1	0	1.06
CBR with 3 clusters	108.6	0.4	24.4	0	-0.16
CBR with 10 clusters	30.29	0.28	21.00	30	13.79
CBR with 8 clusters	31.26	0.35	26.99	12.5	-10.79

We define the CBR system's success rate by the beam angle and beam number errors, thus, the improvement of these values over all cases in the case base denotes the overall improvement of the revised CBR system's performance. That is, the success rate is defined by the percentage of the improvement in average BA and BN error values in the revised CBR system compared to the ones in the original CBR system.

The revised CBR system retrieved treatment plans with decreased isocentre distances between the retrieved treatment plan and known treatment plan of the target case compared to the original CBR system.

The average BA error value has decreased in CBR system with 2 clusters. However, in the CBR system with 3 clusters the BA error has increased by 0.16% compared to the average BA error in the original system. The average BN error value remains the same as in the original CBR system, 0.4, for the revised systems with 2 and 3 clusters. The revised CBR systems with 2 and 3 clusters retrieved the same cases as the original CBR system in 85% of cases. Although the revised CBR system with 3 clusters had better isocentre than the original CBR system the average BA error is worse than the one in the original system.

The revised CBR system with 8 clusters has decreased isocentre distances between retrieved case and target case from 207.87, in the original system, to 31.26 in the revised system. The average BN error has improved by 12.5% from 0.4 to 0.35. However, the BA error value has increased compared to the original CBR system from 24.36 to 26.99. Thus, the lower BN error in the revised CBR system with 8 clusters was obtained at the expense of BA error. That is, cases retrieved treatment plans that suggested better solution with respect to beam number, however, worse solution with respect to beam angles compared to the original CBR system. Thus, the success rate of the revised CBR system is lower by 10.79% with respect to BA error. The revised CBR system with 8 clusters retrieved cases different from the original CBR system in 71% of cases.

The CBR system with 10 clusters performed the best and retrieved cases with the isocentre distance decreased to an average of 30.29. Average BA error value has decreased by 13.79% from 24.36 to 21.0. The average BN error has decreased by 30% from 0.4 to 0.28. The revised CBR system with 10 clusters retrieved cases different to the original CBR system in 80% of cases.

To summarise, the revised CBR system using clustering of cases retrieves cases whose treatment plans have isocentres that match the isocentre of the target cases better than the original system. Although, all of the revised CBR systems have improved isocentre distances compared to the original CBR system, the beam angle error between the retrieved and target cases did not improve in all of them. The improvement in beam angle error was noted in the CBR system with 2 clusters. The CBR system with 3 clusters did not improve the original CBR system. In the CBR system with 8 clusters improvement in beam number error was noted at the expense of beam angle error. The revised CBR system with 10 clusters performed best with respect to both the average beam angle and beam number error. Therefore, we determine 10 clusters to be an optimal number for our data set and use the CBR system with 10 clustering.

6. Conclusion

The revised CBR system with 10 clusters performed best with respect to both the average beam angle and beam number error. The success rate of the revised CBR system with 10 clusters was improved by 30% with respect to the beam number error and by 13.79% with respect to beam angles error compared to the original CBR system. With the retrieved better treatment plan the beam angle adaptation should be easier to conduct.

The treatment planning machine settings have been changed in Nottingham City Hospital, which means that the beam angles in the treatment plans in the case base and the treatment plans of the six new cases obtained from the City Hospital are not directly comparable. In the future we will calibrate these cases and include them in the case base. In addition, the improved CBR system will be presented to medical physicists for further evaluation. Medical physicists adapt treatment plans by placing beams on the patient case image and adjusting them in a trial and error fashion. Our future work will investigate automated ways of treatment plan adaptation in which cases will contain expertise of medical physicists in adaptation of treatment plans. Also, the CBR system needs to be validated, therefore, we will also investigate the possibility of automated validation of the system. Finally, the system will be evaluated by medical physicists.

Acknowledgments

Authors of this paper would like to thank Angela McKenna and Louise Newton and other medical staff at Nottingham City Hospital for providing us with data, their expertise in radiotherapy treatment planning and their valuable time.

References

[1] National Cancer Intelligence Unit 2013. Office for National Statistics, UK (Accessed 02 Nov 2014).
[2] Oldham M, Khoo VS, Rowbottom CG, Bedford JL and Webb S. 1998 A case study comparing the relative benefit of optimizing beam weights, wedge angles, beam orientations and tomotherapy in stereotactic radiotherapy of the brain *Physics in medicine and biology* **43** 21

[3] Shepard D, Ferris M, Olivera G and Mackie T. 1999 Optimizing the delivery of radiation therapy to cancer patients *Siam Review, JSTOR* **41** 721-744
[4] Holder A and Salter B 2004 A tutorial on radiation oncology and optimization *Emerging Methodologies and Applications in Operations Research* (Kluwer Academic Press:Boston)
[5] Rocha H, Dias J, Ferreira B and Lopes M. 2012 Combinatorial optimization for an improved transition from fluence optimization to fluence delivery in IMRT treatment planning *Optimization* **61** 969-987
[6] Kolodner, J. 1992 An introduction to case-based reasoning *Artificial Intelligence Review* **6** 3-34
[7] Jagannathan R, Petrovic S, McKenna A and Newton L 2012 A Novel two phase retrieval mechanism for a clinical case based reasoning system for radiotherapy treatment planning *Int. J. Artif. Intell.* **21** 46-70
[8] Jagannathan R and Petrovic S 2012 A Local Rule-Based Attribute Weighting Scheme for a Case-Based Reasoning System for Radiotherapy Treatment Planning Case-Based Reasoning Research and Development, *Lecture Notes in Computer Science* (Springer: Berlin Heidelberg) **7466** 167-181
[9] MacQueen J 1967 Some methods for classification and analysis of multivariate observations *Proc. Fifth Berkeley Symp. on Math. Statist. and Prob.* (University of California Press:Berkeley, CA) **1** 281-297
[10] Hartigan, J. A. 1975 *Clustering Algorithms* (Wiley: New York)
[11] Magnetoencephalography data. MRI coordinate systems: The head coordinate system (Accessed 02 Nov 2014).
[12] Neuroanatomy Atlas. MRI brain coordinate system 2002 McCausland Center for Brain Imaging, University of South Carolina (Accessed 02 Nov 2014).
[13] Kaufman L and Rousseeuw P J 1990 Partitioning around medoids *Finding groups in data: an introduction to cluster analysis* 68-125
[14] Fraley C and Raftery A E 2002 Model-based clustering, discriminant analysis, and density estimation *Journal of the American Statistical Association* **97** 611-631
[15] Calinski T and Harabasz J 1974 A Dendrite Method for Cluster Analysis *Commun. Stat.* **3** 1-27

Noncoplanar beam angle optimization in IMRT treatment planning using pattern search methods

Humberto Rocha[1], **Joana M Dias**[1,2], **Brígida C Ferreira**[3] **and Maria C Lopes**[4]

[1] INESC-Coimbra, Coimbra, Portugal
[2] Faculdade de Economia, Universidade de Coimbra, Coimbra, Portugal
[3] Escola Superior de Tecnologia de Saúde do Porto, Porto, Portugal
[4] Serviço de Física Médica, IPOC-FG, EPE, Coimbra, Portugal

E-mail: hrocha@mat.uc.pt

Abstract. Radiation therapy is used to treat localized cancers, aiming to deliver a dose of radiation to the tumor volume to sterilize all cancer cells while minimizing the collateral effects on the surrounding healthy organs and tissues. The planning of radiation therapy treatments requires decisions regarding the angles used for radiation incidence, the fluence intensities and, if multileaf collimators are used, the definition of the leaf sequencing. The beam angle optimization problem consists in finding the optimal number and incidence directions of the irradiation beams. The selection of appropriate radiation incidence directions is important for the quality of the treatment. However, the possibility of improving the quality of treatment plans by an optimized selection of the beam incidences is seldom done in the clinical practice. Adding the possibility for noncoplanar incidences is even more rarely used. Nevertheless, the advantage of noncoplanar beams is well known. The optimization of noncoplanar beam incidences may further allow the reduction of the number of beams needed to reach a clinically acceptable plan. In this paper we present the benefits of using pattern search methods for the optimization of the highly non-convex noncoplanar beam angle optimization problem.

1. Introduction

Cancer is one of the most significant health problems worldwide with respect to its incidence and mortality alike. One of the main treatment forms besides surgery and chemotherapy is radiation therapy. Approximately 60% of all patients diagnosed with cancer, which corresponds to nearly 7.6 million patients worldwide, benefit from radiation therapy, either to cure the disease or to palliate symptoms. With this therapy, several beams of ionizing radiation, sent at different incidence directions, pass through the patient. The intersection of these beams is centered at the tumor attempting to sterilize all cancer cells while the surrounding healthy organs and tissues receive radiation from some but not all radiation beams and may thus be spared. An important type of radiation therapy is intensity-modulated radiation therapy (IMRT), a modern technique where the radiation beam is modulated by a multileaf collimator allowing the irradiation of the patient using non-uniform radiation fields from selected angles. Multileaf collimators enable the transformation of the beam into a grid of smaller beamlets of independent intensities. A common way to solve the IMRT optimization problems is to use a beamlet-based approach leading to a large-scale programming problem. Due to the complexity of the whole optimization problem, computation of mathematical algorithms is mandatory to achieve valuable solutions.

Typically, radiation is generated by a linear accelerator mounted on a gantry that can rotate along a central axis and is delivered with the patient immobilized on a movable couch. The movement of the couch combined with the rotation of the gantry allows radiation from almost any angle around the tumor. Despite the fact that almost every angle is possible for radiation delivery, the use of coplanar angles, i.e., angles that lay in the plane of rotation of the gantry, is predominant. Furthermore, regardless of the evidence presented in the literature that selecting appropriate radiation beam incidence directions – beam angle optimization (BAO) problem – can lead to a plan's quality improvement [4, 5, 37], in clinical practice, most of the time, beam directions continue to be either manually selected by the treatment planner in a time-consuming trial and error iterative process or patients are irradiated using evenly spaced coplanar beams. The difficulty of solving the BAO problem, a highly non-convex problem with many local minima on a large search space [14], is one possible justification for the manual selection or the use of coplanar incidence directions. However, evidence shows that noncoplanar incidence directions can lead to better treatment plans compared to plans obtained with coplanar incidence directions [5, 33, 38]. Moreover, the use of noncoplanar incidence directions is decisive for some types of cancers cases, e.g., brain tumors. More recently, some studies have been published where the advantages of noncoplanar incidences have also been shown for extra-cranial treatments like pancreatic cancer [11], prostate cancer [33], mediastinal lymphomas [13], or lung [36]. All of them report a significant increase of organs sparing without losing target coverage or homogeneity.

In this paper we present the benefits of using pattern search methods for the optimization of the highly non-convex noncoplanar BAO problem. A set of clinical cases of head-and-neck tumors treated at the Portuguese Institute of Oncology of Coimbra is used to discuss the potential of this approach in the optimization of the noncoplanar BAO problem. The paper is organized as follows. In the next Section we describe the noncoplanar BAO problem. Pattern search methods framework is presented in Section 3. Computational tests using clinical examples of head-and-neck cases are presented in Section 4. In the last Section we have the conclusions.

2. Noncoplanar beam angle optimization in IMRT treatment planning

The BAO problem is a quite difficult problem to solve since it is a highly non-convex optimization problem with many local minima. In most of the previous works on BAO, the entire range, $[0°, 360°]$ in the coplanar case, of gantry angles is discretized into equally spaced beam directions with a given angle increment, such as 5 or 10 degrees, where exhaustive searches are performed directly or guided by a variety of different heuristics including simulated annealing [7], genetic algorithms [19], particle swarm optimization [24] or other heuristics incorporating a priori knowledge of the problem [20]. Although those global heuristics can theoretically avoid local optima, globally optimal or even clinically better solutions cannot be obtained without a large number of objective function evaluations. The concept of beam's-eye-view (BEV) has been a popular approach to address the BAO problem as well [28]. This approach uses topographic criteria to rank the candidate beam directions. Despite the computational time efficiency of these approaches, the quality of the solutions proposed cannot be guaranteed since the interplay between the selected beam directions is ignored. Other approaches include set cover [22], neighborhood search approaches [3], hybrid approaches [6] or gradient searches [14]. Aleman et al. [4] propose a response surface approach and include noncoplanar angles in beam orientation optimization. Lim and Cao [25] propose an approach that consists of two sequential phases: branch-and-prune and local neighborhood search. Schreibmann et al. [34] propose a hybrid multiobjective evolutionary optimization algorithm for IMRT treatment planning and apply it to the optimization of the number of incident beams, their orientations and intensity profiles. Ehrgott et al. [21] discuss a mathematical framework that unifies the approaches found in literature.

Many of the previous BAO studies are based on a variety of scoring methods or approximations of the fluence map optimization (FMO) problem to gauge the quality of the beam angle set leading to beam angle sets with no guarantee of optimality and questionable reliability since it has been extensively reported that optimal beam angles for IMRT are often non-intuitive [35]. The optimal solution of the FMO problem has been used to drive the coplanar BAO problem [3, 14, 26, 34] including in our works [19, 29, 30, 31]. Many mathematical optimization models and algorithms have been proposed for the FMO problem, including linear models [32], mixed integer linear models [23], nonlinear models [12], and multicriteria models [8]. Our approach for modeling the noncoplanar BAO problem uses the optimal solution value of the FMO problem as the measure of the quality for a given beam angle set. Thus, we will present the formulation of the noncoplanar BAO problem followed by the formulation of the FMO problem we used. We will assume that the number of beam angles is defined a priori by the treatment planner.

2.1. Noncoplanar BAO Model

Let us consider n to be the fixed number of (noncoplanar) beam directions. Let θ denote the gantry angle and ϕ denote the couch angle. Note that, the usual coplanar angles are obtained for a fixed couch position at $\phi = 0$. In our formulation, instead of a discretized sample, all continuous gantry angles and couch angles will be considered. Since the angle $-1°$ is equivalent to the angle $359°$ and the angle $361°$ is the same as the angle $1°$, we can avoid a bounded formulation. For the coplanar optimization, there are no limitations on the choice of candidate directions in the optimization problem. However, for noncoplanar setups, collisions between the patient/couch and the gantry may occur for some candidate directions. In order to keep an unbounded formulation, that information will be embedded in the objective function. A simple formulation for the BAO problem is obtained by selecting an objective function such that the best set of beam angles is obtained for the function's minimum:

$$\begin{aligned} \min \quad & f\Big((\theta_1,\phi_1),\ldots,(\theta_n,\phi_n)\Big) \\ s.t. \quad & \Big((\theta_1,\phi_1),\ldots,(\theta_n,\phi_n)\Big) \in \mathbb{R}^n \times \mathbb{R}^n. \end{aligned} \tag{1}$$

In this work, the objective function $f\Big((\theta_1,\phi_1),\ldots,(\theta_n,\phi_n)\Big)$ that measures the quality of the set of beam directions $(\theta_1,\phi_1),\ldots,(\theta_n,\phi_n)$ corresponds to the optimal value of the FMO problem for each fixed set of beam directions and incorporates the collisions that may occur between the patient/couch and the gantry for some candidate directions:

$$f\Big((\theta_1,\phi_1),\ldots,(\theta_n,\phi_n)\Big) = \begin{cases} +\infty & \text{if collisions occur} \\ \text{optimal value of the FMO} & \text{otherwise.} \end{cases}$$

The used formulation and resolution of the FMO problem are presented next.

2.2. FMO formulation and resolution

Treatment plan optimization is inherently a multicriteria process. However, typically, the FMO problem is modeled as a weighted sum function where constraints are often implemented as objectives. This formulation of the FMO problem makes it harder to capture an accurate trade-off between objectives without violating constraints and may lead to inferior treatment plans. Multicriteria approaches for the FMO problem have been proposed in recent works [15, 27]. The advantage of these approaches is the possibility of selecting, *a posteriori*, a desired solution from a set of Pareto-optimal treatment plans. This option is not suitable to include in a fully automated BAO process. Recently, a multicriteria optimization approach has been developed, where a set of *a priori* defined criteria have to be met (objectives and constraints)

Table 1. Wish-list for head and neck cases.

	Structure	Type	Limit			
Constraints	PTV1	Maximum	74.9 Gy (=107% of prescribed dose PTV1)			
	PTV2	Maximum	63.6 Gy (=107% of prescribed dose PTV2)			
	PTV1 shell	Maximum	59.5 Gy (=85% of prescribed dose PTV1)			
	PTV2 shell	Maximum	50.5 Gy (=85% of prescribed dose PTV2)			
	Spinal cord	Maximum	45.0 Gy			
	Brainstem	Maximum	54.0 Gy			
	Body	Maximum	70.0 Gy			
	Structure	**Type**	**Priority**	**Goal**	**Sufficient**	**Parameters**
Objectives	PTV1	Minimize LTCP	1	1	0.5	T_i = 70 Gy; α= 0.75
	PTV2	Minimize LTCP	2	1	0.5	T_i = 59.4 Gy; α= 0.75
	Parotid (R)	Minimize mean	3	26 Gy	–	–
	Parotid (L)	Minimize mean	4	26 Gy	–	–
	Body	Minimize mean	5	–	–	–

[8, 9, 10]. This approach is suited for an automated BAO process. In this work, a multicriterial optimization based on a prescription called wish-list [8, 9, 10] is our choice to address the FMO problem. The wish-list contains hard constraints and prioritized objectives. Constraints have to be strictly met while objectives are optimized taking into account priorities defined in the wish-list. The higher an objective priority is, the higher the probability that the corresponding objective will be met. The wish-list used for the clinical examples of retrospective treated cases of head and neck tumors at the Portuguese Institute of Oncology of Coimbra is given in Table 1. In general, the head and neck region is a complex area to treat with radiotherapy due to the large number of sensitive organs in this region (e.g. eyes, mandible, larynx, oral cavity, etc.). For simplicity, in this study, the organs at risk (OARs) used for treatment optimization were limited to the spinal cord, the brainstem and the parotid glands. The tumor to be treated plus some safety margins is called planning target volume (PTV). For the head and neck cases in study, two PTVs (PTV1 and PTV2) were defined, to be treated with a simultaneous integrated boost technique. The wish-list contains seven constraints and five objectives based on the prescribed doses for all the structures considered in the optimization. All constraints are maximum-dose constraints. To support the dose optimization in the tumor volumes, PTV1 shell and PTV2 shell are surfaces constructed by computerized volume expansions (ring with no width) positioned at 10 mm distance from PTV1 and PTV2, respectively. The imposed maximum-dose constraints avoid high doses far from the PTVs. Objectives with priorities 1 and 2 aim at dose coverage of PTV1 and PTV2, respectively. Priorities 3–5 aim at sparing the parotids and unspecified tissue (Body).

For the target dose optimization, Breedveld et. al [8, 9, 10] choose the logarithmic tumor control probability (LTCP) [1],

$$LTCP = \frac{1}{N_T} \sum_{i=1}^{N_T} e^{-\alpha(D_i - T_i)},$$

where N_T is the number of voxels in the target structure, D_i is the dose in voxel i, T_i is the prescribed dose, and α is the cell sensitivity parameter. For doses D_i lower than T_i, the LTCP has an exponential penalty while for doses higher than the prescribed dose, the value slowly approaches 0. LTCP equals 1 for a homogeneous dose equal to T_i. A higher α results in less voxels with a low dose, and thus a higher percentage of the PTV receiving 95% of the prescribed dose (PTV coverage).

The algorithm used for multicriterial optimization of beam intensity profiles using the wish-list is the $2p\epsilon c$ [9]. The algorithm is organized in two phases. In the first phase, following the prioritization of the wish-list, the objectives are minimized within constraints. After each objective minimization, a constraint based on the minimized objective results is added to be used in the minimization of the next priority objectives. This action guarantees that higher order priority objective results are not deteriorated by lower priority objective optimization. In the end of the first phase, the plan obtained fulfills all the constraints of the wish-list and each objective has an attained value that is equal to its goal (even if better results could have been achieved), or higher than its goal if the constraints (including higher priority objective constraints) did not allow better results. In the second phase, following the prioritization, all objectives apart from LTCP objectives are sequentially fully minimized. The minimization for the LTCP objectives stops at the defined sufficient value to leave space for lower prioritized objectives improvement, and not to escalate the dose. For a detailed description of the algorithm see Breedveld et al.[9].

The FMO model is used as a black-box function and the conclusions drawn regarding BAO coupled with this formulation/resolution are valid also if different FMO formulations/resolutions are considered.

3. Pattern Search Methods

Pattern search methods are derivative-free optimization methods that require few function value evaluations to converge and have the ability to avoid local entrapment. These two characteristics gathered together make pattern search methods suited to address the BAO problem.

Pattern search methods use the concept of positive bases (or positive spanning sets) to move towards a direction that would produce a function decrease. A positive basis for $\mathbb{R}^n$ can be defined as a set of nonzero vectors of $\mathbb{R}^n$ whose positive combinations span $\mathbb{R}^n$ (positive spanning set), but no proper set does. A positive spanning set contains at least one positive basis. It can be shown that a positive basis for $\mathbb{R}^n$ contains at least $n+1$ vectors and cannot contain more than $2n$ [18]. Positive bases with $n+1$ and $2n$ elements are referred to as minimal and maximal positive basis, respectively. Commonly used minimal and maximal positive bases are $[I \ \ -e]$, with I being the identity matrix of dimension n and $e = [1 \ \ldots \ 1]^\top$, and $[I \ \ -I]$, respectively. The motivation for directional direct search methods such as pattern search methods is given by one of the main features of positive basis (or positive spanning sets) [18]: there is always a vector $\mathbf{v}^i$ in a positive basis (or positive spanning set) that is a descent direction unless the current iterate is at a stationary point, i.e., there is an $\alpha > 0$ such that $f(x^k + \alpha\mathbf{v}^i) < f(x^k)$. This is the core of directional direct search methods and in particular of pattern search methods. The notions and motivations for the use of positive bases, its properties and examples can be found in [2, 18].

Pattern search methods are iterative methods generating a sequence of non-increasing iterates $\{x_k\}$. Given the current iterate x^k, at each iteration k, the next point x^{k+1}, aiming to provide a decrease of the objective function, is chosen from a finite number of candidates on a given mesh M_k defined as

$$M_k = \{x^k + \alpha_k \mathbf{V}\mathbf{z} : \mathbf{z} \in \mathbb{Z}_+^p\},$$

where α_k is the mesh-size (or step-size) parameter, $\mathbb{Z}_+$ is the set of nonnegative integers and $\mathbf{V}$ denote the $n \times p$ matrix whose columns correspond to the p ($\geq n+1$) vectors forming a positive spanning set.

Pattern search methods are organized around two steps at every iteration. The first step consists of a finite search on the mesh, free of rules, with the goal of finding a new iterate that decreases the value of the objective function at the current iterate. This step, called the search step, has the flexibility to use any strategy, method or heuristic, or take advantage of a priori knowledge of the problem at hand, as long as it searches only a finite number of points in the mesh. The search step provides the flexibility for a global search since it allows searches away from the neighborhood of the current iterate, and influences the quality of the local minimizer or stationary point found by the method.

If the search step fails to produce a decrease in the objective function, a second step, called the poll step, is performed around the current iterate. The poll step follows stricter rules and, using the concepts of positive bases, attempts to perform a local search in a mesh neighborhood around $\mathbf{x}^k$, $\mathcal{N}(\mathbf{x}^k) = \{\mathbf{x}^k + \alpha_k \mathbf{v} : \text{ for all } \mathbf{v} \in P_k\} \subset M_k$, where P_k is a positive basis chosen from the finite positive spanning set $\mathbf{V}$. For a sufficiently small mesh-size parameter α_k, the poll step is guaranteed to provide a function reduction, unless the current iterate is at a stationary point [2]. So, if the poll step also fails to produce a function reduction, the mesh-size parameter α_k must be decreased. On the other hand, if both the search and poll steps fail to obtain an improved value for the objective function, the mesh-size parameter is increased or held constant. The most common choice for the mesh-size parameter update is to halve the mesh-size parameter at unsuccessful iterations and to keep it or double it at successful ones.

4. Computational results

The pattern search methods framework was tested for the optimization of the noncoplanar BAO problem using a set of two clinical examples of retrospective treated cases of head-and-neck tumors at the Portuguese Institute of Oncology of Coimbra (IPOC). Treatment plans with five to nine equispaced coplanar beams are used at IPOC and are commonly used in practice to treat head-and-neck cases [3]. We considered plans with five beams because the importance of BAO increases when a lower number of beam directions is considered. Therefore, treatment plans of five coplanar and noncoplanar orientations were obtained using pattern search methods and denoted *coplanar* and *noncoplanar*, respectively. These plans were compared with the typical 5-beam equispaced coplanar treatment plans denoted *equi*. The objective of these comparisons is twofold. First, to demonstrate that pattern search methods are suited to address the highly non-convex BAO problem producing good results both in 2D (coplanar) and 3D (noncoplanar) search spaces. Second, to show that using this approach, higher quality treatment plans using a noncoplanar beam angle set can be obtained.

We choose to implement the pattern search methods algorithm taking advantage of the availability of an existing pattern search methods framework implementation used successfully by us to tackle the BAO problem [29, 30, 31] – the last version of SID-PSM [16, 17]. The spanning set used was the positive spanning set ($[e \ \ -e \ \ I \ \ -I]$). Each of these directions corresponds to, respectively, the rotation of all incidence directions clockwise, the rotation of all incidence directions counter-clockwise, the rotation of each individual incidence direction clockwise, and the rotation of each individual incidence direction counter-clockwise. Since we want to improve the quality of the typical equispaced treatment plans, the starting point considered is the equispaced coplanar 5-beam angle set. To address the BAO problem, efficiency on the number of function value computation is of the utmost importance. Therefore, the number of trial points in the search step should be minimalist. In the last version of SID-PSM, the search step was provided with the use of minimum Frobenius norm quadratic models to be minimized within a trust region, which can can lead to a significant improvement of direct search for smooth, piecewise smooth, and noisy problems [16]. However, these models cannot be computed until a minimum number of points is evaluated. Moreover, the points already tested should span the search space as best as possible to increase the radius of the search. Therefore, both for the coplanar and noncoplanar cases, in the first iterations, the trial points computed in the search step correspond to the points presented in the previous section, that span the 2D and the 3D search spaces, respectively.

Our tests were performed on a Dell Precision T5600 with 8-core Intel Xeon processador E5-2687W, 64GB 1600MHz DDR3 ECC RDIMM. For importing DICOM images, compute and

Journal of Physics: Conference Series **616** (2015) 012014 doi:10.1088/1742-6596/616/1/012014

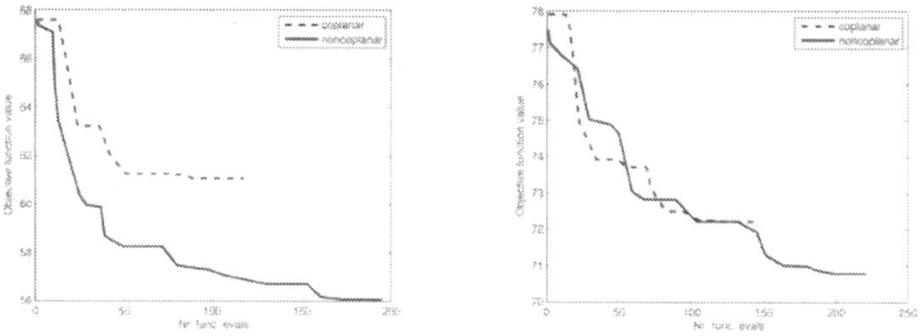

(a) (b)

Figure 1. History of the 5-beam angle optimization process for coplanar and noncoplanar angles, considering the equispaced configuration (equi) as starting point, for cases 1 and 2, 1(a) to 1(b) respectively.

Table 2. Target coverage, conformity and homogeneity obtained by treatment plans.

Case	Target parameters	*noncoplanar*	*coplanar*	*equi*
1	PTV1 Coverage	0.983	0.944	0.943
	PTV1 Conformity	0.140	0.114	0.087
	PTV1 Homogeneity	0.909	0.899	0.901
	PTV2 Coverage	0.940	0.940	0.940
	PTV2 Conformity	0.733	0.710	0.726
	PTV2 Homogeneity	0.793	0.779	0.779
2	PTV1 Coverage	0.994	0.984	0.985
	PTV1 Conformity	0.176	0.202	0.158
	PTV1 Homogeneity	0.921	0.913	0.919
	PTV2 Coverage	0.957	0.952	0.942
	PTV2 Conformity	0.570	0.582	0.598
	PTV2 Homogeneity	0.801	0.795	0.776

visualize dose, and optimize dose distributions, we used YARTOS, an in-house optimization suite developed at Erasmus MC Cancer Institute in Rotterdam. YARTOS is written in MATLAB and contains an optimizer based on a primal-dual interior-point algorithm, capable of solving general nonlinear non-convex optimization problems, and tailored for IMRT treatment planning making full use of multi-threaded computing. The optimal value of the FMO problem used to drive our BAO algorithm was obtained using the YARTOS optimizer.

The history of the 5-beam angle optimization process, presented in terms of the number of function evaluations, is shown in Figure 1 for the two clinical cases of head-and-neck tumors. Both *coplanar* and *noncoplanar* plans obtained considerable improvements in terms of FMO optimal value, with respect to the initial point (typical clinically used equispaced configuration *equi*), with the later obtaining better results at the cost of few more function value evaluations. Despite the improvement in the FMO value, the quality of the results can be perceived considering a variety of metrics. A metric usually used for plan evaluation is the volume of PTV that receives 95% of the prescribed dose (coverage). Typically, 95% of the PTV volume is required. The conformity and homogeneity are other metrics typically screened. These metrics are output values of the YARTOS optimizer and are reported in Table 2. We can

Table 3. OARs sparing obtained by treatment plans.

		Mean Dose (Gy)			Max Dose (Gy)		
Case	OAR	*noncoplanar*	*coplanar*	*equi*	*noncoplanar*	*coplanar*	*equi*
1	Spinal cord	–	–	–	48.311	48.429	49.715
	Brainstem	–	–	–	51.760	54.484	54.694
	Left parotid	17.980	20.082	23.142	–	–	–
	Right parotid	21.253	23.510	26.396	–	–	–
2	Spinal cord	–	–	–	45.233	45.174	45.100
	Brainstem	–	–	–	54.179	54.240	54.629
	Left parotid	25.126	25.367	28.303	–	–	–
	Right parotid	34.844	35.950	38.069	–	–	–

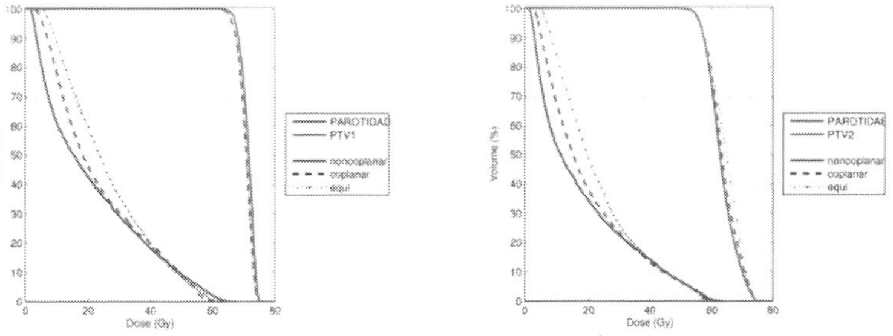

Figure 2. Cumulative dose volume histogram comparing the results obtained by *noncoplanar*, *coplanar* and *equi* for the first case.

verify that *noncoplanar* treatment plans consistently obtained slightly better target coverage, conformity and homogeneity numbers compared to *coplanar* treatment plans. Both optimized plans outperform the *equi* treatment plans. Mean and/or maximum doses of OARs are usually displayed to verify organ sparing. Organ sparing results are shown in Table 3. The maximum dose values for the spinal cord and the brainstem are similar for the different treatment plans, with a slightest advantage for the *noncoplanar* treatment plans. However, as expected, the main differences reside in parotid sparing. The optimized treatment plans enhance better parotid sparing compared to the *equi* treatment plans. The *noncoplanar* treatment plans manage to obtain an average reduction of the parotid's mean dose irradiation in 4.2 Gy compared to the *equi* treatment plans while the *coplanar* treatment plans only obtained a 2.7 Gy average decrease.

Typically, results are judged also by their cumulative dose-volume histogram (DVH). For illustration, DVH results for the first patient are displayed. For clarity, only parotids and tumor volumes are displayed and the DVHs were split as an attempt to better visualize the results. The results displayed in Figure 2 confirm the benefits of using the optimized beam directions obtained and in particular noncoplanar treatment plans.

5. Discussion and conclusions

The BAO problem is a continuous global highly non-convex optimization problem known to be extremely challenging and yet to be solved satisfactorily. An approach for the resolution of the noncoplanar BAO problem, using a pattern search methods framework, was proposed and tested using two clinical head-and-neck cases. Pattern search methods framework is a suitable approach for the resolution of the non-convex BAO problem due to their structure, organized around two phases at every iteration. The poll step, where convergence to a local minima is assured, and the search step, where flexibility is conferred to the method since any strategy can be applied. The use of minimum Frobenius norm quadratic models to be minimized within a trust region are used in the search step and can lead to a significant improvement of direct search for the type of problems at hand, particularly if a prior proper exploration of the search space is guaranteed. Adding to the search step flexibility, and similarly to other derivative-free optimization methods, when minimizing non-convex functions with a large number of local minima, pattern search methods have the ability to avoid being trapped by the closest local minima of the starting iterate, and find a local minima in lowest regions.

For the clinical cases retrospectively tested, the use of noncoplanar directions in our approach showed a positive influence on the quality of the minimizer found. The improvement of the local solutions in terms of objective function value corresponded, for the head-and-neck cases tested, to high quality treatment plans with better target coverage and with improved organ sparing, in particular better parotid sparing. Moreover, we have to highlight the low number of function evaluations required to obtain locally optimal solutions, even using noncoplanar directions and consequently exploring a much larger search space, which is a major advantage compared to other global heuristics where the continuous solution space is confined to a discrete subset, often only considering coplanar irradiation angles to achieve clinically acceptable computation times. The efficiency on the number of function value computations is of the utmost importance, particularly when the BAO problem is modeled using the optimal values of the FMO problem.

Acknowledgements

We would like to show gratitude to Ben Heijmen and Sebastiaan Breedveld for giving us permission to install Yartos. This work was supported by QREN under Mais Centro (CENTRO-07-0224-FEDER-002003) and FEDER funds through the COMPETE program and Portuguese funds through FCT under project grant PTDC/EIA-CCO/121450/2010. This work has also been partially supported by FCT under project grant PEst-OE/EEI/UI308/2014.

References

[1] M. Alber, R. Reemtsen, Intensity modulated radiotherapy treatment planning by use of a barrier-penalty multiplier method, Optim. Methods Software 22 (2007) 391–411.

[2] P. Alberto, F. Nogueira, H. Rocha, L.N. Vicente, Pattern search methods for user-provided points: Application to molecular geometry problems, SIAM J. Optim. 14 (2004) 1216–1236.

[3] D. M. Aleman, A. Kumar, R. K. Ahuja, H. E. Romeijn, J. F. Dempsey, Neighborhood search approaches to beam orientation optimization in intensity modulated radiation therapy treatment planning, J. Global Optim. 42 (2008) 587–607.

[4] D. M. Aleman, H. Romeijn, J. Dempsey, A response surface approach to beam orientation optimization in intensity modulated radiation therapy treatment planning, INFORMS J. Comput. 21 (2009) 62–76.

[5] M. Bangert, P. Ziegenhein, U. Oelfke, Comparison of beam angle selection strategies for intracranial IMRT, Med. Phys. 40 (2013).

[6] D. Bertsimas, V. Cacchiani, D. Craft, O. Nohadani, A hybrid approach to beam angle optimization in intensity-modulated radiation therapy, Comput. Oper. Res. 40 (2013) 2187–2197.

[7] T. Bortfeld, W. Schlegel, Optimization of beam orientations in radiation therapy: some theoretical considerations, Phys. Med. Biol. 38 (1993) 291–304.

[8] S. Breedveld, P. Storchi, M. Keijzer, A.W. Heemink, B. Heijmen, A novel approach to multi-criteria inverse planning for IMRT, Phys. Med. Biol. 52 (2007) 6339–6353.

[9] S. Breedveld, P. Storchi, B. Heijmen, The equivalence of multicriteria methods for radiotherapy plan optimization, Phys. Med. Biol. 54 (2009) 7199–7209.

[10] S. Breedveld, P. Storchi, P. Voet, B. Heijmen, iCycle: integrated, multicriterial beam angle, and profile optimization for generation of coplanar and noncoplanar IMRT plans, Med. Phys. 39 (2012) 951–963.

[11] D.S. Chang, G.K. Bartlett, I.J. Das, H.R. Cardenes, Beam angle selection for intensity-modulated radiotherapy (IMRT) treatment of unresectable pancreatic cancer: are noncoplanar beam angles necessary?, Clin. Transl. Oncol. 15 (2013) 720–724.

[12] K. Cheong, T. Suh, H. Romeijn, J. Li, J. Dempsey, Fast Nonlinear Optimization with Simple Bounds for IMRT Planning, Med. Phys. 32 (2005) 1975–1976.
[13] X. Chen, D. Jin, S. Wang, M. Li, P. Huang, J. Daia, Noncoplanar intensity-modulated radiation therapy for young female patients with mediastinal lymphoma, J. Appl. Clin. Med. Phys. 13 (2012) 147–156.
[14] D. Craft, Local beam angle optimization with linear programming and gradient search, Phys. Med. Biol. 52 (2007) 127–135.
[15] D. Craft, T. Halabi, H. Shih, T. Bortfeld, Approximating convex Pareto surfaces in multiobjective radiotherapy planning, Med. Phys. 33 (2006) 3399–3407.
[16] A. L. Custódio, H. Rocha, L. N. Vicente, Incorporating minimum Frobenius norm models in direct search, Comput. Optim. Appl. 46 (2010) 265–278.
[17] A. L. Custódio, L. N. Vicente, Using sampling and simplex derivatives in pattern search methods, SIAM J. Optim. 18 (2007) 537–555.
[18] C. Davis, Theory of positive linear dependence, Am. J. Math. 76 (1954) 733–746.
[19] J. Dias, H. Rocha, B.C. Ferreira, M.C. Lopes, A genetic algorithm with neural network fitness function evaluation for IMRT beam angle optimization, Cent. Eur. J. Oper. Res. 22 (2014) 431–455.
[20] W.D. D'Souza, R.R. Meyer, L. Shi, Selection of beam orientations in intensity-modulated radiation therapy using single beam indices and integer programming, Phys. Med. Biol. 49 (2004) 3465-3481.
[21] M. Ehrgott, A. Holder, J. Reese, Beam Selection in Radiotherapy Design, Linear Algebra and its Applications 428 (2008) 1272–1312.
[22] C-H. Lee, D.M. Aleman, M.B. Sharpe, A set cover approach to fast beam orientation optimization in intensity modulated radiation therapy for total marrow irradiation, Phys. Med. Biol. 56 (2011) 5679–5695.
[23] E. K. Lee, T. Fox, I. Crocker, Integer programing applied to intensity-modulated radiation therapy treatment planning, Ann. Oper. Res. 119 (2003) 165–181.
[24] Y. Li, D. Yao, J. Yao, W. Chen, A particle swarm optimization algorithm for beam angle selection in intensity modulated radiotherapy planning, Phys. Med. Biol. 50 (2005) 3491–3514.
[25] G.J. Lim, W. Cao, A two-phase method for selecting IMRT treatment beam angles: Branch-and-Prune and local neighborhood search, Eur. J. Oper. Res. 217 (2012) 609–618.
[26] V.V. Misic, D.M. Aleman, M.B. Sharpe, Neighborhood search approaches to non-coplanar beam orientation optimization for total marrow irradiation using IMRT, Eur. J. Oper. Res. 205 (2010) 522–527.
[27] M. Monz, K. H. Kufer, T. R. Bortfeld, C. Thieke, Pareto navigation Algorithmic foundation of interactive multi-criteria IMRT planning, Phys. Med. Biol. 53 (2008) 985–998.
[28] A. Pugachev, L. Xing, Pseudo beam's-eye-view as applied to beam orientation selection in intensity-modulated radiation therapy, Int. J. Radiat. Oncol. Biol. Phys. 51 (2001) 1361-1370.
[29] H. Rocha, J.M. Dias, B.C. Ferreira, M.C. Lopes, Beam angle optimization for intensity-modulated radiation therapy using a guided pattern search method, Phys. Med. Biol. 58 (2013) 2939–2953.
[30] H. Rocha, J.M. Dias, B.C. Ferreira, M.C. Lopes, Selection of intensity modulated radiation therapy treatment beam directions using radial basis functions within a pattern search methods framework, J. Glob. Optim. 57 (2013) 1065–1089.
[31] H. Rocha, J.M. Dias, B.C. Ferreira, M.C. Lopes, Pattern search methods framework for beam angle optimization in radiotherapy design, Appl. Math. Comput. 219 (2013) 10853–10865.
[32] H.E. Romeijn, R.K. Ahuja, J.F. Dempsey, A. Kumar, A column generation approach to radiation therapy treatment planning using aperture modulation, SIAM J. Optim. 15 (2005) 838–862.
[33] L. Rossi, S. Breedveld, B. Heijmen, P. Voet, N. Lanconelli, A. Shafak, On the beam direction search space in computerized non-coplanar beam angle optimization for IMRT-prostate SBRT, Phys. Med. Biol. 57 (2012) 5441-5458.
[34] E. Schreibmann, M. Lahanas, L. Xing, D. Baltas, Multiobjective evolutionary optimization of the number of beams, their orientations and weights for intensity-modulated radiation therapy, Phys. Med. Biol. 49 (2004) 747–770.
[35] J. Stein, R. Mohan, X.H. Wang, T. Bortfeld, Q. Wu, K. Preiser, C.C. Ling, W. Schlegel, Number and orientation of beams in intensity-modulated radiation treatments, Med. Phys. 24 (1997) 149–160.
[36] G.M.M. Videtic, K. Stephans, C. Reddy, S. Gajdos, M. Kolar, E. Clouser, T. Djemil, Intensity-Modulated Radiotherapy-Based Stereotactic Body Radiotherapy for Medically Inoperable Early-Stage Lung Cancer: Excellent Local Control, Int. J. Radiat. Oncol. Biol. Phys. 77 (2010) 344–349.
[37] P. Voet, S. Breedveld, M. Dirkx, P. Levendag, B. Heijmen, Integrated multicriterial optimization of beam angles and intensity profiles for coplanar and noncoplanar head and neck IMRT and implications for VMAT, Med. Phys. 39 (2012) 4858–4865.
[38] X. Wang, X. Zhang, L. Dong, H. Liu, M. Gillin, A. Ahmad, K. Ang, R. Mohan, Effectiveness of noncoplanar IMRT planning using a parallelized multiresolution beam angle optimization method for paranasal sinus carcinoma, Int. J. Radiat. Oncol. Biol. Phys. 63 (2005) 594–601.

System design and improvement of an emergency department using Simulation-Based Multi-Objective Optimization

A Goienetxea Uriarte[1], **E Ruiz Zúñiga, M Urenda Moris and A H C Ng**

School of Engineering Science, University of Skövde, PO Box 408, Skövde, SE 54128 Sweden

E-mail: ainhoa.goienetxea@his.se

Abstract. Discrete Event Simulation (DES) is nowadays widely used to support decision makers in system analysis and improvement. However, the use of simulation for improving stochastic logistic processes is not common among healthcare providers. The process of improving healthcare systems involves the necessity to deal with trade-off optimal solutions that take into consideration a multiple number of variables and objectives. Complementing DES with Multi-Objective Optimization (SMO) creates a superior base for finding these solutions and in consequence, facilitates the decision-making process. This paper presents how SMO has been applied for system improvement analysis in a Swedish Emergency Department (ED). A significant number of input variables, constraints and objectives were considered when defining the optimization problem. As a result of the project, the decision makers were provided with a range of optimal solutions which reduces considerably the length of stay and waiting times for the ED patients. SMO has proved to be an appropriate technique to support healthcare system design and improvement processes. A key factor for the success of this project has been the involvement and engagement of the stakeholders during the whole process.

1. Introduction

Healthcare facilities and specially Emergency Departments (ED) are complex systems to design and operate, some of the main reasons are: the high amount of different resources involved in the processes, the uncertainty of these processes occurring at different moments and the high possibility of simultaneously needed resources [1]. As a result, long patient waiting times are a usual problem on EDs. In order to better understand the inherent complexity and improve the system, the use of different simulation methodologies is applied. One of the most commonly used simulation method is Discrete Event Simulation (DES). Additionally, considering the multiple objectives that are typically involved in the design and improvement of an ED the need to apply a Simulation-based Multi-Objective Optimization (SMO) approach is superior to a single objective method.

This article presents the results of an ongoing research project developed by the Production and Automation Engineering Division of the University of Skövde together with the ED of the regional Skaraborg Hospital Skövde (SkaS) in Sweden. The aim of this paper is to present the process followed to improve the ED of SkaS in reaching the required target patient waiting times and length of stay

[1] To whom any correspondence should be addressed.

defined for EDs in Sweden. The article describes the steps from the development of a DES model to the final stage of applying SMO for decision support.

The structure of the article is divided as follows: a literature review is presented in the second section; followed by a description of the ED project; the fourth section describes the results of the optimization; and to finalize, some conclusions and future work are identified.

2. Literature review

EDs are becoming the main entrance of patients to hospitals. It is reported by many institutions that around 50% of the patients admitted to the hospital come from the ED [2]. Moreover, emergency care units are characterized by their complexity due to the stochastic behaviour of patient arrivals and the unpredictable care needed by them [3], as well as the occupation of shared staff and resources between the ED and correlated departments [4]. Additionally, an ED is usually intended to deal with patients in a critical or life threatening situation instead of dealing with patients presenting low acuity injuries or illnesses [3] which makes the system even more complex. A specific issue that also characterizes EDs is the reception of patients. The first evaluation of the patient is made in the triage area where his condition and acuity level is defined, generally by a nurse. Many EDs have a poorly conceived first encounter system; usually the triage areas are too small or without enough capacity to face the stochastic pattern of patient's arrivals [2]. This common situation combined with the priority sorting due to patient´s acuity, results in long waiting times for patients, acting as a bottleneck of the system [2].

Simulation is an analytical tool to create, maintain, evaluate or improve a system or process. The first use of simulation within healthcare systems dates back to the fifties. It was applied to increase the efficiency in the use of resources of a healthcare unit and later on, simulation was identified as the most powerful tool for healthcare system analysis and improvement [5].Since the early 1990s the number of studies applying simulation in healthcare has grown rapidly [6]. Nowadays, the application of simulation technologies for the improvement of healthcare systems has increased considerably [4, 7]. The main use of simulation in ED systems has been the utilization of the so called "what-if" scenarios. Within these scenarios new patient arrival patterns, resource settings and work proceedings can be tried out without disturbing the real system, or be developed even before a system is constructed [8]. An important benefit of simulation when it comes to improve systems is that it helps to identify the bottlenecks of the process. In healthcare systems this means there are not enough beds, resources or staff to treat the patients efficiently from the arrival to the discharge [8].

DES is necessary in order to be able to model and represent the complex and stochastic flow of patients that usually go through healthcare clinics [9]. Furthermore, it is the most reported technique for healthcare improvement in the literature followed by Montecarlo Simulation and System Dynamics [1]. There are several simulation studies that present how DES can be used to identify improvements or better design EDs. Gunal and Pidd [3] analyse how to increase the ED performance through DES and Hay et al. [4] present a new ED modelling approach. Ferrin et al. use DES to improve the flow and access to care, but the main outcome was to demonstrate simulation´s unique ability to target parameters to maximize the operational and financial impact [8]. Additional studies look at improving patient flow and throughput analysis [10-13] and estimating the future capacity of the ED [14, 15].

DES provides the results of specific what-if experiment scenarios. But in order to analyse several scenarios, a large amount of modelling time is usually required and although an improved solution can be found, the optimal one is not guaranteed. Since simulation is not an optimization tool by itself, a step that combines simulation and optimization is needed [16]. Traditionally, simulation and optimization have been considered as different approaches in the operational research domain, but they have developed together and finally the idea is to use the great detail of simulation together with the ability of optimization to give optimal solutions [17]. It has been demonstrated that combining optimization and simulation tools allow decision-makers to quickly determine optimal system configurations, even for complex integrated facilities [9]. Depending on the type of problem to analyse, there are different optimization methods that can be used in combination with simulation, of which several are presented by Figueira and Almada-Lobo [17]. Meta-heuristic optimization is a flexible approach to examine any solution space and it is characterized by quickly achieving good quality solutions, therefore it has usually been used in combination with DES [17]. Consequently, the integration of meta-heuristic optimization together with simulation is necessary if the optimal range of solutions for the given input is wanted. Moreover, if there are multiple objectives to be analysed at the same time then SMO is the correct approach to be applied. SMO facilitates the search for trade-offs between several conflicting objectives [18].

Different authors have described how simulation and optimization techniques applied together have improved EDs, for example Ahmed and Alkhamis [19] use Monte Carlo simulation together with genetic algorithms for optimal staff allocation. Chen [20] presents in his article an analysis of an ED where SMO was applied to find an optimal ED resource allocation. Nevertheless, there are not many cases in the literature where DES and SMO have been applied together to analyse EDs.

3. Emergency Department study

3.1. Background

In the Swedish healthcare system, the responsibility for health and medical care is shared by the National Board of Health and Welfare (SoS), county councils and municipalities. Due to the long patient waiting times that are experienced in EDs in the country, the SoS set out specific targets that should be met by all the county councils in Sweden. These targets were the following: 90% of the patients should get a triage within 10 minutes from arrival; 90% of patients should meet a physician within 1 hour from arrival; and 90% of patients should have a maximum total length of stay of 4 hours. Despite the efforts done in previous years to reach these targets, the ED unit at SkaS was not making significant improvements towards reaching them. The waiting times were far from what was expected. For example, the time to triage was in some cases up to one hour and the time to meet a physician and the length of stay of patients were also far from the established objectives. Considering the situation, SkaS contacted the Production and Automation Division at the University of Skövde in order to extend their collaboration and target the ED and its objectives.

3.2. Emergency department and process description

SkaS is the largest hospital in the area of Skaraborg in Sweden, giving service to around 277.000 inhabitants. The ED is open 24 hours a day and receives an average of 51.000 visits per year. Besides its internal capacity, the ED shares resources and staff with other departments of the hospital, such as the children department, the X-ray department and the laboratory unit. The ED is divided in 4 different specialities according to the patient classification: surgery, orthopaedics, medicine and children. The personnel is composed by resident and intern physicians for every speciality of the ED, registered nurses (RN), ambulance nurses, triage nurses, laboratory and X-ray personnel and the receptionist. The resources which have been considered and modelled are: the reception, the waiting room, the triage rooms, the surgery area, the orthopaedics area, the medicine area, the children area, the laboratory, the X-Ray department, the urgent care rooms and the patient observation areas.

The process begins when a patient enters the ED. There are two paths for patients who arrive to the ED, walk-in (66.6% of the patients) and ambulance patients (33.3%). The ambulance patients usually get a triage in the ambulance and once they arrive to the ED they are redirected to an emergency care room, area speciality room or to the observation area, depending on the acuity and type of care needed. The walk-in patients wait in the waiting room until they are transferred to a triage room. The RNs conduct the first examination of the patient. The necessary samples are taken, the care priority is established, the necessary documentation is filled and the routine each patient needs before he is seen by a physician is established. After this triage process, the patient is registered and sent back to the waiting room or to a designated speciality room (surgery, orthopaedics, medicine or children). Once the patient has arrived to the room, he waits to be examined by a physician and thereafter to be sent at home or to take additional laboratory samplers or X-Ray scans (in the case of orthopaedic patients).

Journal of Physics: Conference Series **616** (2015) 012015 doi:10.1088/1742-6596/616/1/012015

The patient stays in the room until the results are received and evaluated by the physician. Lab results are often ready within one hour and in the case of X-ray or scans, patients have to visit (usually accompanied by a nurse) the X-ray unit of the hospital. If additional tests are needed, the same procedure would follow i.e. waiting for results and meeting the doctor (for modelling purposes, a maximum of three meetings with the doctor has been estimated). When the process at the ED has finished, the doctor sends the patient, either home or to a ward of the hospital. Finally, the required documentation about the patient is filled in by the physician (considered administrative task in the model).

3.3. Modelling the Emergency Department

The modelling process that has been followed is based on the steps defined by Banks et al.[21]. It started by defining the objectives for the project and ensuring a clear understanding of the process. This was done through iterative meetings and discussions with the stakeholders of the hospital and several visits to the ED. This step was followed by a considerable data collection and analysis phase. The stakeholders provided data from an entire year containing around 50.000 patient visits to the ED. Information about patient arrival time and duration of the different activities performed during his stay, as well as doctor and nurses´ activities were extracted from the data. Non-reliable data, such as wrong patient registration or records with missing information, were excluded from the data analysis. A few number of processes such as the time to get the results from the laboratory or the time the doctors spend with the patients in each meeting were missing, consequently a time study was conducted in order to get this information. The results of the time study were validated with the ED personnel before using them into the simulation model.

The number of patients' arrivals each month, day and hour is different and also the acuity level and type of patient can vary. The number of patients arriving each hour was spaced randomly in the model within that hour. In order to imitate the behaviour of the real system, the data was analysed in detail and different probabilities and statistical distributions were defined in the model accordingly. Additionally, different parameters such as variable service time for patients, variable number (1-3) and duration of meetings with the physician depending on the type of patient and its acuity, priority waiting time queue depending on the acuity and the length of stay of the patient, variable pattern of visits to the X-Ray department, variable administrative task for physicians, variable waiting time for lab and X-Ray processes, etc. were introduced.

A number of assumptions were made and implemented in the model. These assumptions were defined together with the hospital personnel in order to limit the complexity of the system or when needed data was missing, but also to ensure that the results would not be compromised in any case.

Taking into account the data described above, a detailed DES simulation model was implemented in FlexSim HC 3D Simulation Software. The main resources simulated were the patients, rooms and beds, resident and intern physicians (surgery, medicine, orthopaedics and children), the receptionist, triage nurses, ambulance nurses and RNs (surgery, medicine, orthopaedics and children), laboratory and X-Ray staff. Resident physicians are considered expert physicians who spend a percentage of their time interrupted either by hospital calls or assisting other departments of the hospital and also supporting intern physicians in the ED. Interns have been considered to be slower than the residents, spending an additional 20% of the estimated time when meeting a patient. The number of nurses has been modelled according to the reality but their activities haven´t been introduced in detail, this will be done in a future project.

The model was verified and validated in order to ensure that its behaviour is an accurate representation of the real-world system [22]. A simulation of 90 days and 10 replications was run in order to verify and validate the model. The validation phase consisted in analysing the average and 90 percentile of the Time To Triage (TTT), Time To first Meeting with the Doctor (TMD) and Length Of Stay (LOS) for all the ED patients. A maximum deviation of 5% for the average and 10% for the 90 percentile was established (there are some exceptions for low values of specific type of patients). The implication of the stakeholders was a key factor in this stage of the process. Various meetings were held with them in order to validate the output and adjust parameters if needed.

In table 1 the LOS values (mean and 90 percentile values) are presented for the total amount of patients, ambulance patients and walk-in patients. The simulation results were compared with the data obtained from the real system. This comparison was also performed for eight sub-classifications of patients in order to validate the model accurately. The columns under "difference" show the variance between the results of the real system and the model.

Table 1. Model validation results. Length of stay.

	Real ED (minutes)		Difference (%)		Model (minutes)	
	P90	Mean	P90	Mean	P90	Mean
Total patients	265.00	144.69	3.95	0.07	276.30	144.80
Ambulance patients	263.00	146.25	6.20	-2.11	279.30	143.16
Walk-in patients	267.00	143.96	3.10	0.97	275.29	145.36

The TTT table, table 2, represents the time from the patient arrival (just for walk-in patients) until the patient enters the triage room. In this case, comparing the difference column, there is a value bigger than the established maximum of 5 %. As explained above, this value has been considered as correct due to the low values which are being compared (between 7 and 8 minutes).

Table 2. Model validation results. Time to triage.

	Real ED (minutes)		Difference (%)		Model (minutes)	
	P90	Mean	P90	Mean	P90	Mean
Walk-in patients	21.00	7.85	8.45	-8.36	22.77	7.20

The TMD table, table 3, shows a summary of the validation results of the time patients wait before they meet a physician.

Table 3. Model validation results. Time to first Meeting with the physician.

	Real ED (minutes)		Difference (%)		Model (minutes)	
	P90	Mean	P90	Mean	P90	Mean
Total patients	141.00	63.77	-4.01	2.39	135.34	65.30
Ambulance patients	109.00	46.48	1.59	-0.57	110.74	46.22
Walk-in patients	151.00	72.05	-5.99	0.69	141.95	72.55

The figures from table 1-3 show that the model of the ED of SkaS represents in a fairly accurate manner the real system.

3.4. Designing improvement scenarios

As described above, one of the strengths of DES is the support it offers to design alternative what-if scenarios. The aim of these improvement scenarios is to find a more optimal configuration of the ED in order to achieve the goals defined by the SoS to reduce the patient waiting times and LOS and thereby, increase its service level.

The following table 4 presents a summary of the overall results of the different improvement scenarios. The scenarios are organized in two columns: scenarios without a significant impact and those with a significant improvement impact in the system output.

Table 4. Summary of tested scenarios.

Scenarios without significant impact in the results	Scenarios with significant impact in the results
- To keep opened the X-Ray department 24 hours a day.	- Reduce in 50% the time needed for the X-Ray process.
- Every physician at the ED is modelled as resident physician.	- Reduce in 50% the time needed for the lab process. *(Objective LOS achieved)*
- To eliminate the exits of surgery physicians to the hospital.	- To eliminate physician's exits to the hospital.
- Reduce in 50% the exits to the hospital for all the physicians.	- To open the second triage room when more than two patients are waiting for triage. *(Objective TTT achieved)*

The results from these initial what-if studies show that several system changes lead to a clear system improvement by increasing its service level and efficiency. The waiting time and LOS for some categories of patients was reduced in a significant manner and it lead to an overall system reduction, even achieving the goals defined by the SoS (requirements are on system level). However, it was not possible to get a system solution which reached the goal of the 90% of the patients meeting a physician within 60 minutes from the arrival (TMD).

Combinations of several improved scenarios, personnel and resources were later on implemented in the model without success. The results were close to fulfil the goal of TMD but it implied a disproportionate amount of extra physicians, rooms and process time reductions.

The analysis of the what-if scenarios concluded that the process time for X-Ray and Laboratory were key parameters in order to get good results. The stakeholders were aware of this importance, but the analysis of the scenarios showed that the improvement of just these parameters, although giving better results than the original system, were not enough to achieve the defined goals. Additionally, an implemented scenario solved the TTT objective without the need of having a second triage room opened 24 hours a day. The definition and experimentation stage of these scenarios, took a considerable amount of time and made it still very difficult to get an overview of the impact each combination of parameters and policy options had. Instead of continuing defining new possible scenarios, which would have been a nearly never ending process, the project experimentation effort changed the focus in order to embrace a SMO analysis for the improvement of SkaS ED.

4. Optimization results

The objective of the optimization is to find an optimal configuration of the ED which reaches the TMD and LOS levels defined by the SoS for the different categories of patients (medicine, surgery, orthopaedics and children).

The parameters of the experimentation phase which had a major impact on the system performance, were selected as the input variables of the optimization. These parameters are the amount of different extra physicians (medicine, surgery, orthopaedics and children), the amount of extra beds per department (medicine-children and surgery-orthopaedics) and the reduction of the processing time of three activities: the administrative task performed by physicians, the respond time from the laboratory and the respond time from the X-Ray. In table 5 the range for each one of these optimization input variables is presented.

Table 5. Optimization parameters.

Parameters	Values
X-Ray time reduction	0,5,10,15,20 (minutes)
Laboratory time reduction	0,5,10,15 (minutes)
Administrative task reduction	0,5,10,15 (minutes)
Medicine-children extra beds	0-3
Surgery-orthopaedic extra beds	0-3
Children extra doctors	0-3
Medicine extra doctors	0-3
Orthopaedic extra doctors	0-3
Surgery extra doctors	0-3

The TTT was excluded from the optimization analysis because the changes in the described parameters don´t affect it and a what-if scenario had already found a good enough solution.

Apart from the TMD and LOS, three other objectives were included in order to drive the optimization towards resource effective solutions: to *minimize the total amount of extra physicians*, to *minimize the number of extra beds* and to *minimize the number of reductions* of the different processing times for *physician administration task, respond time for Laboratory* and *respond time for X-Ray*. Without the formulation of these additional objectives, the optimization algorithm would have just selected solutions with a high number of resources without considering its effective use. On the other hand, even though it is important to understand the impact a process time reduction has on the system performance, it is also difficult to implement that change in the real system. The implementation of parameter reduction changes in the real system will require modifications in the working methods and will include not only new technical systems but also to run a Lean standardization process. Method improvement and sustaining the day-to-day system performance are two common approaches to increase and maintain the productivity of a system. The other objectives add new resources (physicians or hospital beds) and represent the typical approach to increase capacity by investing. These parameters are discrete and the implementation of them is much more straight forward and do not really change the way the work is done. It is obviously an easier approach, if physicians are available and if money is not an issue. Sometimes it is even the necessary approach, especially when the system has clear bottlenecks or its capacity is far from the target goals of the business, but from a productivity point of view it should not be the first option.

The chosen optimization algorithm was the NSGA-II, which is a robust multi-objective algorithm that proves to be suitable for this project´s optimization characteristics [23].

The first results from the optimization are presented in the two following Scatter 3D charts, figure 1 and figure 2. Each point in these charts represent a possible configuration of the ED. Figure 1 shows the TMD of patients (TTL) in the X axis, the total amount of physicians (Docs) in the Y axis and the total amount of process time reductions (Reductions) in the Z axis. Figure 2 shows the LOS of patients in the X axis, the total amount of physicians (Docs) in the Y axis and the total amount of process time reductions (Reductions) in the Z axis. In both charts the Pareto front of solutions is shown highlighted. The Pareto front is the set of optimal solutions for the different combinations of personnel, resources, waiting times and processing times.

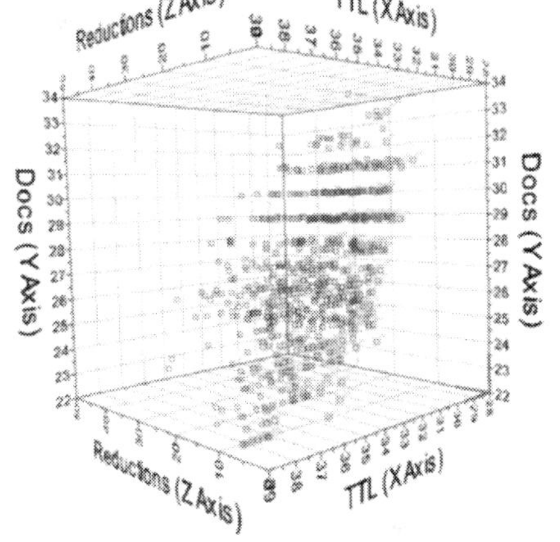

Figure 1. Optimization results. Scatter 3D chart of TMD (TTL), number of physicians (Docs) and process time reductions (Reductions).

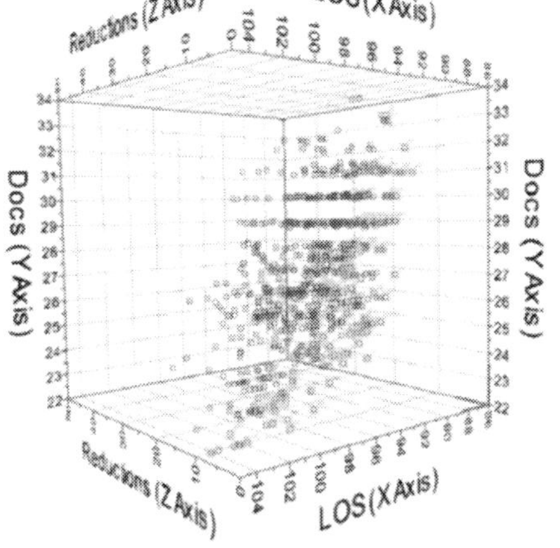

Figure 2. Optimization results. Scatter 3D chart of LOS, number of physicians (Docs) and process time reductions (Reductions).

The following chart in figure 3 represents around 1000 possible combinations of the ED of SkaS with the given inputs. Every combination shows its correspondent average values of the TMD (TTL in the chart) and LOS of patients. The optimization parameters (see table 5) are placed in the initial columns and the results of LOS and TMD in the last two columns. The numbers in each column correspond to the values of each parameter (amount of minutes and the number of beds or doctors depending on the parameter). The colours in the diagram represent a range of values in the resultant LOS or TMD (the range for TMD is shown in the chart below).

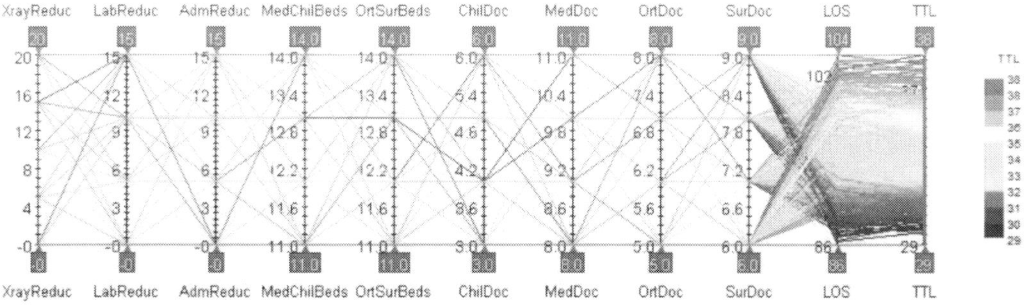

Figure 3. Parallel coordinates chart. Optimization results of LOS and TMD (TTL in the chart) in minutes. The rest of parameters shown in columns are in minutes or number of resources.

This parallel coordinates chart is used to help decision makers navigate among the possible optimal trade-off solutions generated by the optimization. With this chart they can establish maximum and minimum values for each variable /column in order to define possible solutions in areas that are more interesting. It is not uncommon that the initial budget or objectives shifts during a project. Having access to the whole solution space makes it easier for decision makers to look for solutions that are relevant to the new circumstances. Another important parameter that would impact in the final solution is the availability of physicians. The final solution can therefore be optimal but still tailor made to the category and number of physicians that the clinic manages to recruit. All these considerations can easily be included in the final solution thanks to the chosen multi-objective approach.

An example of some specific optimal solutions is shown in figure 4. The boundaries of the input parameters have been delimited and as a result four optimal solutions are shown. These solutions include adding to the ED a maximum of one extra orthopaedic physician, two extra surgery physicians, two extra surgery-orthopaedic beds, reducing the time for laboratory respond by 10 minutes and 5 minutes for X-Ray respond time.

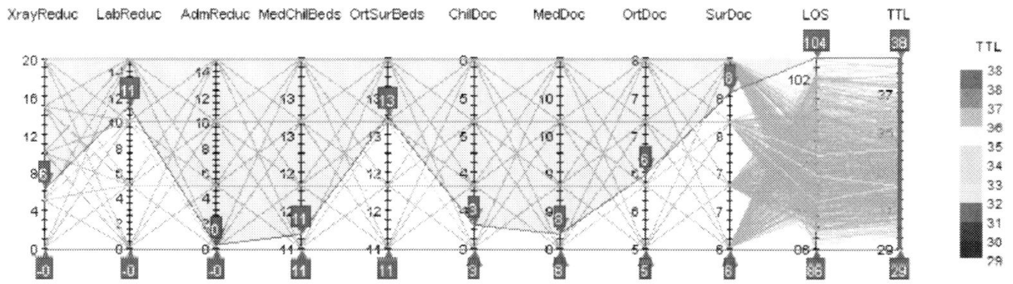

Figure 4. Optimal solutions example.

In the following table 6, the results of one of these possible solutions is presented. It shows the reduction of approximately 45% in the mean value and 90 percentile of TMD and a 35-40% reduction in the mean value and 90 percentile of LOS results compared to the original model.

Table 6. Optimization results. Time to meet the physician and LOS.

	Model (minutes)		Difference (%)		Solution (minutes)	
	P90	Mean	P90	Mean	P90	Mean
Total patients TMD	135.34	65.30	-45.05	-44.40	74.37	36.30
Total patients LOS	276.30	144.80	-38.91	-36.02	168.80	92.65

In the previous table it is possible to appreciate a significant improvement of the TMD and LOS of this particular solution. To implement this solution or a similar one with the mentioned extra physicians, beds and process time reductions would give the ED the chance to get much closer or even achieve the aims defined by the SoS. These reductions mean that if this solution is applied at the real ED, patients would wait 30 minutes less in average to meet a physician (60 minutes for the 90% of the patients) and they would have a 50 minutes shorter LOS in average (105 minutes for the 90% of the patients).

Analysing the whole set of results provided by the optimization, it can be concluded that for the given optimization parameters and values, the objective of the TMD for the 90% of the patients was not possible to achieve at the levels established by the SoS. Anyway, the optimal solutions show significant possible reductions of around 50% for that specific objective. Additional suggestions from the personnel at the ED and the interaction between SMO and Lean improvement will be considered in a future work at hand in order to get the desired levels.

5. Conclusion and future work

This work shows that SMO can efficiently be used to improve healthcare systems and EDs in particular. The study presents how a number of variables need to be simultaneously improved in order to reach a significant system improvement. Additionally, both the importance and limitations of what-if analysis are described. Furthermore, it highlights the importance and strength of SMO, by being able to truly support decision-making through presenting all optimal trade-off solutions. These solutions have guided the managers at SkaS to know what to do and what not to do without prior testing in the real system. It has given them a much higher insight about their system performance and peculiarities, increasing their knowledge and guiding their work in finding better ways of improving it. This process has saved considerable time, money and resources to the ED and is the foundation to reduce the patient's waiting times and length of stay. Although lots of possible improved combinations have been found as a result of the optimization, not all the objectives defined by the SoS have been achieved yet. Further optimizations with some other configurations of the optimization problem as well as different working methodologies of the ED are now tested in order to get the required results.

The implication and engagement of the stakeholders of the hospital has been a key success factor for the correct development of the project and to define new future collaboration opportunities. In this way, future projects have been identified such as to include a detailed working procedure of the nurses into the model and to participate in the design of the future ED of SkaS.

References

[1] Young T, Eatock J, Jahangirian M, Naseer A and Lilford R 2009 Three critical challenges for modeling and simulation in healthcare. In: *2009 Winter Simulation Conference, WSC 2009,* (Austin, TX, USA: IEEE) pp 1823-30

[2] Harrell C R, Ghosh B K and Bowden R 2011 *Simulation using ProModel* (USA: McGraw Hill Higher Education)

[3] Gunal M M and Pidd M 2006 Understanding accident and emergency department performance using simulation. In: *2006 Winter Simulation Conference, WSC,* (Monterey, CA, USA: IEEE) pp 446-52
[4] Hay A M, Valentin E C and Bijlsma R A 2006 Modeling emergency care in hospitals: A paradox - The patient should not drive the process. In: *2006 Winter Simulation Conference, WSC,* (Monterey, CA, USA: IEEE) pp 439-45
[5] Denney S H 1997 MedModel - healthcare simulation software. In: *Proceedings of the 1997 Winter Simulation Conference,* ed S Andradottir, *et al.* (Piscataway, NJ, United States Atlanta, GA, USA: IEEE) pp 581-6
[6] Brailsford S C, Harper P R, Patel B and Pitt M 2009 An analysis of the academic literature on simulation and modelling in health care *J. Simul.* **3** 130-40
[7] Roberts S D 2011 Tutorial on the simulation of healthcare systems. In: *2011 Winter Simulation Conference, WSC 2011,* (Phoenix, AZ, USA: IEEE) pp 1403-14
[8] Ferrin D M, Miller M J and McBroom D L 2007 Maximizing hospital finanacial impact and emergency department throughput with simulation. In: *2007 Winter Simulation Conference, WSC,* (Washington, DC, USA: IEEE) pp 1566-73
[9] Jacobson S H, Hall S N and Swisher J R 2006 *Patient Flow: Reducing Delay in Healthcare Delivery* (Los Angeles: Springer US)
[10] Blasak R E, Starks D W, Armel W S and Hayduk M C 2003 The use of simulation to evaluate hospital operations between the emergency department and a medical telemetry unit. In: *Proceedings of the 2003 Simulation Conference: Driving Innovation,* ed S E Chick, *et al.* (New Orleans, LA, USA: IEEE) pp 1887-93
[11] Samaha S, Armel W S and Starks D W 2003 The use of simulation to reduce the length of stay in an emergency department. In: *Proceedings of the 2003 Simulation Conference: Driving Innovation,* ed S E Chick, *et al.* (New Orleans, LA, USA: IEEE) pp 1907-11
[12] Mahapatra S, Koelling C P, Patvivatsiri L, Fraticelli B, Eitel D and Grove L 2003 Pairing Emergency Severity Index5-level triage data with computer aided system design to improve emergency department access and throughput. In: *Proceedings of the 2003 Simulation Conference: Driving Innovation,* ed S E Chick, *et al.* (New Orleans, LA, USA: IEEE) pp 1917-25
[13] Takakuwa S and Shiozaki H 2004 Functional analysis for operating emergency department of a general hospital. In: *Proceedings of the 2004 Winter Simulation Conference,* ed R G Ingalls, *et al.* (Washington, DC, USA: IEEE) pp 2003-11
[14] Baesler F F, Jahnsen H E and DaCosta M 2003 The use of simulation and design of experiments for estimating maximum capacity in an emergency room. In: *Proceedings of the 2003 Simulation Conference: Driving Innovation,* ed S E Chick, *et al.* (New Orleans, LA, USA: IEEE) pp 1903-6
[15] Wiinamaki A and Dronzek R 2003 Using simulation in the architectural concept phase of an emergency department design. In: *Proceedings of the 2003 Simulation Conference: Driving Innovation,* ed S E Chick, *et al.* (New Orleans, LA, USA: IEEE) pp 1912-6
[16] AlDurgham M M and Barghash M A 2008 A generalised framework for simulation-based decision support for manufacturing *Production Planning and Control* **19** 518-34
[17] Figueira G and Almada-Lobo B 2014 Hybrid simulation-optimization methods: A taxonomy and discussion *Simulation Modelling Practice and Theory* **46** 118-34
[18] Deb K 2001 *Multi-objective optimization using evolutionary algorithms* (Wiltshire, UK: John Wiley & Sons, LTD)
[19] Ahmed M A and Alkhamis T M 2009 Simulation optimization for an emergency department healthcare unit in Kuwait *European Journal of Operational Research* **198** 936-42
[20] Chen T L 2014 Decision support system based on distributed simulation optimization for medical resource allocation in emergency department. In: *Lecture Notes in Computer Science*

(including subseries Lecture Notes in Artificial Intelligence and Lecture Notes in Bioinformatics), pp 15-24

[21] Banks J, Carson J S, Nelson B L and Nicol D M 2000 *Discrete-Event Simulation* (Upper Saddle River, New Jersey: Prentice Hall, Inc.)

[22] Banks J 1998 *Handbook of Simulation: Principles, Methodology, Advances, Applications, and Practice* (New York, USA: John Wiley & Sons, Inc.)

[23] Deb K, Pratap A, Agarwal S and Meyarivan T 2002 A fast and elitist multiobjective genetic algorithm: NSGA-II *IEEE Transactions on Evolutionary Computation* **6** 182-97

Simulation of "a week care unit" organization in an academic hospital

G. Sinnaeve [1], C. Beguin [2], P. Chevalier [3] and M. Philippe [4]

[1,3] Université Catholique de Louvain, Voie du Roman Pays, 34, L1.03.01 - B-1348 Louvain-la-Neuve, Belgium
[2,4] Cliniques universitaires Saint-Luc, Avenue Hippocrate 10 - 1200 Bruxelles, Belgium

E-mail: gilles.sinnaeve@uclouvain.be

E-mail: claire.beguin@uclouvain.be

E-mail: philippe.chevalier@uclouvain.be

E-mail: marianne.philippe@uclouvain.be

Abstract. Poor bed allocation to each specialty might lead to a surplus of available beds during weekends. The introduction of a "week unit" that is only available during working days may improve the efficiency of bed allocation by uniting the patient flow of two regular units, before converting the regular beds of one unit into week beds. In order to test the possible introduction of this system in three services of the Saint-Luc hospital, we created a simulation model that considers new allocation rules to direct the patients into the right unit.

1. Introduction

Historically, the organization of care units has always been based on a predetermined number of beds allocated to each medical specialty. This distribution method leads to an inefficient use of beds when considering patient flow. A surplus in the number of available beds during weekends is particularly common. A strategy that combines one regular hospitalization unit with one "week unit" that is only available during working days may improve the efficiency of bed allocation.

1.1. Different types of SSU

Short stay units (SSUs) take a variety of forms in the literature, depending on the author's definition. One generally accepted basic definition is suggested by Damiani [8]: "It is a type of admission that can provide targeted care for patients requiring brief hospitalization ($<$ 5 days), as well as patients ready to be discharged as soon as their clinical condition is resolved. Usually, the SSU is open from Monday to Friday, 24 hours a day, and provides the same level of medical care as an ordinary ward". In this system, admissions and discharges are specifically timed, with, for example, opening days restricted to week days. Consequently, patient treatment slots are reserved for during the week. In contrast to Damiani, other authors have proposed alternative definitions based on location, access, and opening hours. The majority considers an SSU to be closely linked to the emergency department [1]. Its access can be restricted to an explicit list of pathologies, such as respiratory diseases [2, 3], or to thyroidectomy patients [4]. It can also be solely dedicated to patients in good enough condition to be likely to recover within a short period of time, i.e., with a lower rate of complications [5, 6]. In addition, opening times may range from 5 to 7 days a week.

1.2. The week unit and its advantages
SSUs offer many advantages that have been widely documented in the literature. Firstly, the nursing staff has increased capacity to deal with rapid cases, depending on the SSU's specialization. Secondly, the entire team becomes more and more accustomed to the related administrative tasks. As a result of these two advantages, many authors [8] have observed a reduction in the average length of stay, without any side-effects. The quality of care provided has been shown to remain stable or improve, while the same number of patients still receive treatment during the week days. The hospital can also save on human cost during the weekends, when at least eight beds are converted into week beds [9]. Reviewing bed allocations for each medical and surgical specialty of a hospital could, however, prove a complex task [10]. An SSU could be introduced to compensate for poor bed utilization rates during the weekends that are the result of the current organization of the operating area.

1.3. Discrete event simulation
Discrete event simulation is becoming an increasingly popular tool in the healthcare field. It allows us to replicate a real-world system in order to virtually experiment with some changes. The consequences of the changes are gathered and analyzed before being implemented in a real setting [11]. The system's performance is evaluated on appraisal of how the selected performance indicators evolve when faced with modifications in some parameters. "What if" scenarios are developed and compared in order to determine an ad hoc solution based on the elements from the real system. For an overview of the potential applications of discrete event simulation (DES) in the healthcare sector, we refer to Jacobson [12]. The steps and skills required to build a valid simulation model were developed in full by Carson [13] and White in 2009 [14].

1.4. Contribution of the paper
In this paper, we will henceforth refer to a specific type of SSU that only concerns hospitalization units open 5 days a week. This so-called "week unit" implies uniting the patient flow of two regular units, before converting the regular beds of one unit into week beds. The bed allocation is then organized based on the estimated length of stay, with short-stay patients allocated to the week unit and long- or varying-stay patients to the regular unit. In cases where short-stay patients have to be admitted for a weekend, they are transferred to the regular unit. This definition is compatible with that developed by Renaut in 2009 [9]. To the best of our knowledge, the week unit system has never before been virtually tested.

We believe that our study could contribute by:

(i) Further increasing the growing number of simulation applications in the healthcare field;
(ii) Shedding light on the performance of a week unit;
(iii) Evaluating the impact of introducing a week unit at a university hospital.

2. Methods
Configuring a week unit consists of separating the patient flows based on new allocation rules. These rules are based on specific criteria designed to predict patient length of stay. In this way, patients from the gynecology, urology, and ophthalmology services are directed either to the week or regular unit. To select the best length of stay predictors, we needed to analyze and compare each criterion. To this end, we accumulated a database including all hospitalizations in 2011, with data on personal information and surgical operations. We could then use this historical data to build our simulation model, rendering it capable of conducting tests to compare how well

the two kinds of patients were separated. The preferred algorithm was one exhibiting the lowest number of inter- or extra-unit transfers and the most balanced utilization.

2.1. The Saint-Luc hospital and its services
Founded in 1976, Saint-Luc is an academic hospital in Brussels equipped with 979 beds. These beds are distributed across the 10 floors of the hospitalization tower. Each floor is composed of four units of care, with each unit offering approximately 30 beds. The beds are allocated to rooms that can either be private or double.

We tested the week unit concept in the following three services, spread over two parallel units: gynecology (U94), urology, and ophthalmology (U93). This particular hospital's structure highly favors inter-unit interaction due to the multiple access points linking the different units. These units were chosen for several reasons:

(i) The similarities in the way patients are treated in the urologic and gynecologic services;
(ii) The low proportion of emergency cases;
(iii) The large proportion of short-stay patients;
(iv) The current restricted opening times of the U94 (from Monday 7am to Saturday 12am).

2.2. Data collection
The dataset used is the outcome of merging operating databases with the RHM (*Résumé Hospitalier Minimum*, patient information that every Belgian hospital is legally required to submit to the government to qualify for funding). The Belgian financing system relies on rates fixed in advance for each APR DRG (All Patient Refined Diagnosis Related Groups). This classification system uses the previous system of DRGs (Diagnosis Related Groups), which are divided into four different SOIs (Severity Of Illness) [15]. This dataset contains different criteria that were selected and gathered for their ability to predict patient length of stay. Database refers mainly to personal data, specific stay (including DRG, SOI and hospital access) or performed surgical operations. We focused our interest on the data taken from 2011. According to key members of the Saint-Luc hospital team, 2011 presented no specifically unusual data and thus could be representative for an analysis of the three concerned services. Nevertheless, approximately 100 patients who had more complex records were excluded from this study.

2.3. Profile analysis
The database comprises approximately 3,089 stays, divided among three services. There were 1,379 in gynecology (44.64%), 1,070 in urology (34.64%), and 640 in ophthalmology (20.72%) in 2011. Table 1 reveals the most relevant figures of this 2011 sample profile analysis.

Table 1. Basic statistics for the most relevant criteria of the 3 services

Criteria	Gynecology	Urology	Ophthalmology
Proportion of emergency patients	7%	21%	8%
Proportion of weekend patients	25%	51%	45%
Average length of stay (days)	2.75	3.50	2.56
Proportion of stays < 2 nights	54%	51%	78%
Number of DRGs	11	18	8
1 DRG includes > 50% of the patients	DRG 513	None	DRG 71

Basic statistics and comparisons indicate that the DRGs represented the best length of stay predictors among the selected criteria. Nevertheless, their efficiency contrasted with the high number of DRGs per service (Table 1) and the length of stay variability within the groups. Introducing SOIs into the model reduced this variability. However, most of the pathologies treated in these services exhibited a low SOI of 1 or 2, rarely scoring higher.

Finding good predictors of length of stay is reputed to be a challenging task. Most researchers have attempted this for economic purposes [16]. Still, even if DRGs did not appear to be perfect indicators of length of stay, neither personal data nor information concerning potential operation offered better predictive power.

2.4. Model description

A model is an abstraction of a more complex phenomenon. It allows us to simplify the different elements in order to focus our interest on the interactions between selected elements. A model is created in order to simulate the behavior of a process [14]. The AnyLogic software was chosen for this purpose.

In our AnyLogic model, the sources input patients at a predefined rate. Once input, each entity continued through the intermediate stage where patients were assigned to their unit based on an allocation algorithm. This algorithm relied on the DRGs and the SOIs assigned to each patient, as well as their arrival day. One of the aims of this study consisted in selecting the allocation algorithm that would minimize the number of patient transfers while providing the best workload balance within the two units. It decided where to send the patients itself, whereas the actual separation occurred in the switch. If there were no beds available, a weekday patient would first be directed to the regular unit, while a regular patient would be ejected from the system. This model also predefined the processing time, i.e., the time a patient will stay in the unit of care.

The outcome measures chosen for evaluating the system's performance were:

- Utilization rate: the number of patients allocated to each unit ;
- Internal transfers, from the week unit to the regular one ;
- External transfers, from the regular unit to the outside world.

We also considered the case where the units were combined. This model allowed us to test the consequences of considering the bed capacity of the two units together. If some patients were unable to find a bed in this model then even the most perfect allocation algorithm would not be able to achieve a better result.

The hospital stay of each patient from the database was simulated based on the historical data from 2011. In total, 3,089 gynecology, urology, and ophthalmology patients were allocated to one of the 51 available beds of the units 93 and 94. There are two types of rooms, either single (called "private") or double patients. 22 rooms of these units are double patients, while 7 are private. Even if double patients rooms could be privatized, we considered this number of private rooms as fixed. This configuration implies to take into account the separation of men and women in double rooms. We exclude this constraint from the model. Indeed, gender separation should not have a critical impact on the results and can be performed easily during the validation step.

The simulation relies on historical data taken from 2011, for both arrivals and length of stay. Data are mostly reliable on the days of arrivals, but not necessarily on the hours of arrivals and departures, indicated by the nurses when they have free time. This is the reason why we allowed recourse actions to add some flexibility to the simulation.

These recourse actions are presented under "what if" scenarios, and concerned:

(i) The closing time of the week unit;
(ii) The maximum waiting time for a bed when a patient arrive in the system;
(iii) The time left after a stay at the week unit;
(iv) The number of beds available in the regular and week units.

The time left after a stay in the week unit indicates how relevant a transfer from the week unit to the regular one is. Some constraints that were neglected in the virtual model are going to be taken into account in the validation phase. Two extra months will be tested manually in order to confirm the results found in the virtual simulation. This step was required by the hospital staff prior to testing the model in a real context.

2.5. Software description
We used AnyLogic version 6.0 to model the week unit system. This software is widely used in the literature, offering an extensive range of applications. Brailsford *et al.* [24] used this software to build a model combining discrete event simulation (DES) and agent-based modeling (ABM).

3. Results
3.1. The "As-Is" Situation
We started by representing the 2011 unit organization. The patients were directed to their units depending on the service in charge, with gynecology patients directed to unit 94 (U94) and urology and ophthalmology patients to unit 93 (U93). At the time of writing, four beds were dedicated to urology patients in unit 94 due to the rising activity of this service. The U94 closing time typically fixed at Saturday 12pm. In this setting, an average of 10 transfers per month was necessary to sustain the activity.

Table 2. Results for the As-Is situation

Waiting Time (h)	Internal Transfers	External Transfers	U94 patients (%))
2	117	4	45.29

3.2. One pooled unit simulation
This scenario consisted in considering the two units as a pool of 51 beds. The objective was for both teams to be mixed and versatile enough to cope with the variety of diseases from the three services. The U94 closing time was here advanced to Friday 10pm. The operating area organization, as well as the doctor's schedule, remained unchanged. Depending on the simulation, switching to a week unit implied five external transfers, directly related to the early closure of the week unit. The low number of transfers proved the advantage of the week unit system: if the perfect criteria were found, this system could be sustainable for all units concerned.

3.3. What if scenarios with parameters modifications
We selected appropriate indicators that enabled the scenarios to be compared. As a reminder, the routing algorithm was chosen for its ability to direct patients into their respective units while minimizing the number of transfers (internal or external) and providing the best workload balance within the two units. By gathering opinions from the hospital experts, we should be able to find the scenario that best represents their constrained system.

The results derived from the considered scenarios are summarized on table 3 below:

Table 3. Results for the different scenarios

Scenarios	Attempt	1	2	3	4	5	6
Parameters	Closing time (h)	20	21	22	20	20	20
	Time left (h)	12	12	12	12	12	15
	Waiting Time (h)	2	2	2	2	3	2
	Bed capacity	51	51	51	50	51	51
Results	Internal transfers	166	163	160	185	165	142
	External transfers	8	8	8	17	8	6
	Week patients (%)	50.31	50.28	50.31	51.25	50.44	50.28

The first three simulations analyzed the impact of closing the unit 1 or 2 hours later. All other things being equal, this had a limited impact on the number of transfers (internal or external) as well as the workload, measured based on the proportion of patients treated in the week unit. The maximum waiting time allowed for patients to obtain a free bed also has a limited impact. When this time was increased to 3 hours, only one transfer was saved compared to the first scenario, used as a baseline scenario. The parameters that most strongly influenced the indicators were the number of beds and time left for recovery after being transferred.

4. Validation

4.1. Physical validation

Physical validation consists in using a manual model, *i.e.*, post it notes. The relevant information for each patient is recorded on the note, such as identification number, arrival day and time, departure day and time, room, DRG, and service. Physical validation presents several advantages:

(i) Increasing the persuasive power of the model;
(ii) Integrating constraints that are hard to simulate virtually;
(iii) Validating the results.

The health care teams called for a physical validation for two specific months. September 2013 and March 2014 were thus selected for their representative quality and higher number of patients, with 260 and 296 patients respectively.

September 2013 : The system of constraints for this physical simulation has been designed in collaboration with the medical staff. Firstly, doctors wanted to include the type of room (double or private) as a key factor for planning bed allocation. Room type is important to doctors for financial reasons, while patients care about their comfort and privacy. Assigning a double patients room for treating a single patient is now the prior bed allocation rule. Besides, gender separation can now be taken into account. Secondly, the doctors requested the addition of an extra constraint: gynecology and urology patients are not allowed to share a double patients room.

March 2014 : This simulation applied the same constraints as that of September 2013, except for the gynecology and urology room separation. The type of room remained the primary focus, followed by the allocation algorithm and gender room separation rule.

Table 4. Results for the different scenarios

Scenario	Attempt	September 2013	March 2014
Parameters	Closing time (h)	20	20
	Waiting Time (h)	2	2
	Bed capacity	51	51
Results	Internal transfers	7	12
	External transfers	0	3

Flexible bed allocation helps to increase private bed availability. However, when the number of privatized double rooms exceeded a limit on busier days, this was shown to lead to external transfers. This is why three patients were subjected to external transfers in March 2014, in addition to the nine internal ones. These results confirmed those from the simulations. With an average of 13 patients transferred every month over one year, the allocation algorithm based on DRGs provided a fair representation of what could happen in reality. As shown in Appendix B, the number of patients transferred may vary from 4 to 24 transfers per month, with no specific correlation to the number of patient arrivals.

4.2. Field implementation

The health care team reaction of both units was mixed at the announcement of the experiment, initially planned for 6 weeks. This is not surprising considering the size of these teams (about 20 nursing staff members and 10 doctors). After a clear definition of the concept and multiple validations using historical data, it was decided to advance the closing time without changing the allocation system. Ophthalmology and urology patients are still directed to unit 93 while gynecology patients are taken care of at unit 94. The priority is set to avoid any useless transfer. Slowly, doctors are collaborating by adapting the operating area schedule to improve the new discharge policy. Others are making arrangements to transfer their patients into what they consider to be a more suitable unit.

In order to slowly switch to a week unit system, the following steps are advised:

(i) Standardize the processes;
(ii) Improve the collaboration between nursing staff members;
(iii) Anticipating weekly flows variations;
(iv) Investigate on more apt predictors of the length of stay;
(v) Train a single bed allocation manager to coordinate the three services.

The following conclusions may be drawn from the outcome of the 6-week test. Firstly, the two nursing staff teams found a lot of common ground in terms of their working methods. The test also highlighted the need for standardization, especially in terms of administrative processes. Secondly, nurses and bed allocation manager were found to start organizing meetings to plan long stays and forecasted transfers in the clear goal to avoid any useless or damaging transfers. The test revealed that four transfers a week seemed a reasonable workload for the staff.

Finally, Saint-Luc hospital should greatly focus on its communication strategies. The units involved in the test received poor institutional communication about the new system in place, leading to misunderstandings and pressure on both staff and patients. The hospital management decided to extend the test period in order to give enough time to the nursing staff members to get used to this new configuration.

5. Discussion

5.1. Week unit efficiency

We selected the scenario of 51 beds and 12 hours of stay for transferred patients, with a total number of transfers per month approaching approximately 13. The number of transfers, calculated using an allocation algorithm, were compensated by the economic benefits to be gained. Firstly, the increase in the utilization rate, which is difficult to translate into funding, and secondly, the savings made in terms of working time, estimated at 1 ETP per year. According to a similar study in France, savings can be made even after converting just 8 beds into week unit beds [9]. A staff specialized in dealing with short stays could also potentially aid in speeding up the recovery process, reducing the length of stay of week unit patients with no additional risks.

Other strategies could be put into practice in order to improve the efficiency of the week unit. Could doctors organize a gathering to evaluate the patient health status to avoid any useless transfer on Friday afternoon? Switching to a week unit system could also offer a good opportunity to revise current planning practices. Since a unit has to treat the same amount of patients in a shorter time, standardizing and reserving slots for both consultations and operations should enable staff to treat patients without damaging the level of care provided [9]. Schedule reorganization could be combined with a coherent discharge policy in order to reduce bed utilization rate variations [22]. It should not be forgotten that these results were found to induce minimal disturbance to the ward organization and quality of given care.

The model appeared to provide plausible results. We can thus conclude that the week unit system can reasonably be applied in the field.

5.2. Data availability

We observed that the workload was well balanced, with over 50% of patients directed to the week unit, implying an average of 13 transfers per month. This performance was achieved with the DRG system, mainly designed for economic purposes. Theoretically, a better allocation would lead to better performance. Saint-Luc hospital could either use its own internal nomenclature or the medical expertise of the doctors during consultations in order to better predict the length of stay. If this indicator were reviewed periodically, it could improve the predictive accuracy of the model.

5.3. Model

Length of stay and arrivals are both influenced by a large number of parameters that might be correlated, such as bed utilization rate, doctor's availability, season's of the year, DRGs, limited capacity of the Operating Area, etc.

In his attempt to simulate arrival rates, Harrison [18] observed that a subtle seasonal effect could have a significant impact on the bed occupancy rate. However, it is hardly noticeable on the plot of the cumulative arrivals over time (see Figure A). The seasonal effect, combined with LoS variations, implies a variation in the number of monthly transfers, as shown in Appendix C. Here we can see that a high number of monthly transfers did not necessarily correlate to arrival peaks. The plot on Appendix A clearly shows the length of stay variation following the day of arrival for all three specialties combined. By inputting the actual operational data

directly into the system, we were able to avoid making assumptions on the arrival distributions for both emergency and planned arrivals. Poisson distribution, widely used within the literature to simulate arrival rates, corresponded to our data according to classic χ^2 tests conducted in previous studies on the subject [19].

Furthermore, the use of historical data seemed to provide a solid foundation to this simulation, considering the unclear shapes of the length of stay distributions by DRG. The most popular distributions chosen were the LogNormal, Beta, and Exponential distributions. Some authors have proven the superiority of the LogNormal distribution in the context of LoS modeling [20].

A more accurate modelling would imply a much higher cost in data requirement. With the historical data from 2011 only, we succeeded in showing what would happen in 2011 if this system was put into practice. This simulation results from the trade-off between complexity and

5.4. Bed utilization rate
One of the primary objectives to achieve in order to balance the workload was to improve bed utilization rates. We computed the cumulative length of stay of the three services, giving 9,184 hospitalized days in total. With a unit being open approximately 48 weeks per year, we roughly estimated the total opening days for both units at 15,086, prior to introducing the week unit, and 14,186 after. This led to a 4% increase in the average bed utilization rate (from 61 to 65%). Managing the same number of patients a year in 5 days instead of 6 may also raise questions concerning the adequacy of planning organization and activity. Once a stable bed utilization rate is achieved, the staff could use it to better manage the unit.

5.5. Resistance to change
So far, the week unit has only partially been implemented on the field. The service based allocation rule is still effective, nurses only try to decrease the number of transfers while closing their unit one day earlier. This current organization implies an unbalanced workload, leading to possible dissatisfaction. So far, we noticed a poor activity level at unit 94 starting from Friday afternoon. Facing a huge resistance to change from the nursing staff members, the full project still need to be broken down into small steps and require long term assistance in order to be fully effective.

6. Limitation

6.1. Whole-hospital model
Modelling an entire hospital is difficult and costly. According to Cochran in 2006, few are the articles that use Discrete Event Simulation in order to study all hospital interactions. "Most models do not explicitly consider the interactions between different hospital units" [10]. Nevertheless, this model might be adapted individually to see the impact of the week unit system in other units. We might consider modelling units that seem appropriate for week units, and then linking them to take in charge the external transfers. For examples of whole hospital modelling, please refer to Demeester [23] and Schmidt [21], who are both using an affinity matrix to assign patients into wards but differ by their way of handling the patient's length of stay. Closing one unit during the weekends would add complexity to their model but seems reallisticaly feasible.

6.2. Software
AnyLogic is used for its capacity to handle complex interactions between elements. The more elements added to the model, the harder it becomes to add constraints reflecting reality. Like other modelers [24], we experienced limitations in the software beyond its basic functions. Java coding skills were required to overcome this problem. For example, two loops were necessary to limit the length of stay of the week patients following their entry day and time, since they were not allowed to stay in the week unit after Friday 8pm. Nevertheless, this software proved a suitable choice for dealing with a model of this size and level of complexity. For further information about package comparisons with other commercial software available on the market, please refer to [25].

7. Conclusion

This study demonstrates the feasibility of gathering two units into a larger pool, allowing the "week unit" to close during weekends. The optimization of bed allocation paves the way for developing new perspectives in terms of care unit management.

Appendix A.

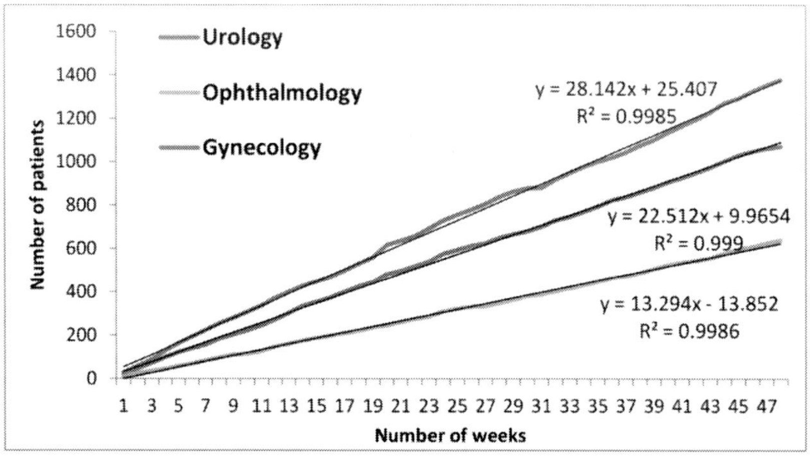

Figure A. Cumulative arrival rate over time by service

Appendix B.

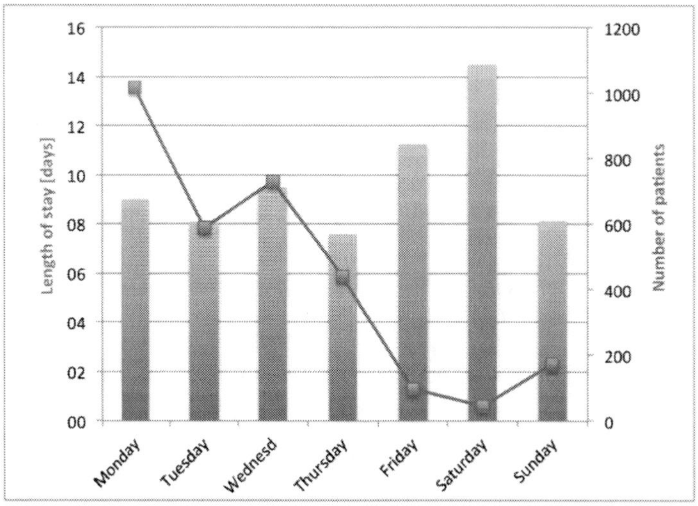

Figure B. Length of stay and patients arrival variations by day of the week

Appendix C.

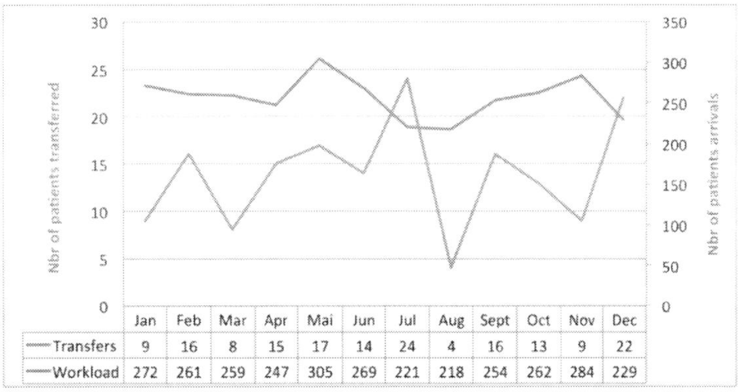

Figure C. Transfers and arrivals by month of the year

References

[1] Najaf-Zadeh A, Hue V, Bonnel-Mortuaire C, Dubos F, Pruvost I and Martinot A 2011 Effectiveness of multifunction paediatric short-stay units: a French multicentre study *Acta Pædiatrica* **100** p. e227–33.
[2] Broquetas J M, Pedreny R, Martínez-Llorens J M, Sellarés J and Geaa J 2008 Short-stay respiratory unit: a new option for inpatient care *Arch Bronconeumol.* **44** 5, p.252–6.
[3] Juan A, Salazar A, Alvarez A, Perez J R, Garcia L and X Corbella 2006 Effectiveness and safety of an emergency department short-stay unit as an alternative to standard inpatient hospitalisation *Emerg Med J* **23** p.833–7
[4] Vrabec S, Oltmann S C, Clark N, Chen H and Sippel R S 2013 A short-stay unit for thyroidectomy patients increases discharge efficiency *Journal of surgical research.***184**, p.104–8.
[5] Abenhaim H A, Kahn S R, Raffoul J and Becker M R 2000 Program description: a hospitalist-run, medical short-stay unit in a teaching hospital *CMAJ.* **163** 11
[6] Downing H, Scott C and Kelly C 2008 Evaluation of a dedicated short-stay unit for acute medical admissions *Clinical medicine.* **8** 1.
[7] O'Brien P, O'Connell C, Fenwick S, Stewart B, Marshall A C and Hickey P 2011 Improved bed use with creation of a short-stay unit in a cardiac catheterization recovery room *Heart and lung.* 40 p.56–62.
[8] Damiani G, Pinnarelli L, Sommella L, Vena V, Magrini P and Ricciardi W 2011 The short stay unit as a new option for hospitals: A review of the scientific literature *Med Sci Monit.* **17** 6, p.15–19.
[9] Renaut L 2009 La performance des structures d'hospitalisation de semaine *Journal de gestion et d'économie médicales.* **27** 5, p.291-295.
[10] Cochran J K and Bharti A 2006 Stochastic bed balancing of an obstetrics hospital *Health care Manage Sci***9** p.31-45.
[11] Hamrock E, Hopkins J, Paige K, Parks J, Scheulen J and Scott Levin 2013 Discrete Event Simulation for Healthcare Organizations: Discrete event simulation for healthcare organizations: a tool for decision making Journal of healthcare management. **58** 2, p.110-124.
[12] Jacobson S H, Hall S N and Swisher J R 2006 Chapter 8: discrete-event simulation of health care systems. patient flow: reducing delay in healthcare delivery in international series *Operations Research Management Science.* **91** p.211–252.
[13] Carson J S 2004 Introduction to modeling and simulation *Winter simulation conference.*
[14] White K P and Ingalls R G 2009 Introduction to simulation *Proceedings of the 2009 Winter Simulation Conference* p.13-23.
[15] Averill R, Goldfield N, Steinbeck B and Grant T 1995 All Patient Refined Diagnosis Related Groups *3M Health Information Systems* Wallingford, Connecticut.
[16] Tan S S, Hakkaart L, van Roijen B, van Ineveld M and Redekop W K 2013 Explaining length of stay variation of episodes of care in the Netherlands *Eur J Health Econ.* **14** p.919-927.
[17] Holm L B, Lurås H and Dahl F A 2013 Improving hospital bed utilisation through simulation and optimisation with application to a 40% increase in patient volume in a Norwegian general hospital *International journal of medical informatics.* **82** p.80-89 .
[18] Harrison W H, Shafer A and Makay M 2005 Modelling Variability in Hospital Bed Occupancy *Health Care Management Science.* **8** p.325-334.
[19] Sinnaeve G, Beguin C, Chevalier P and Philippe M Mémoire sur les unités de semaine *UCL.*
[20] Marazzi A, Paccaud F, Ruffieux C and Beguin C 1998 Fitting the Distributions of Length of Stay by Parametric Models *Medical Care* **36** 6 p.915-927.
[21] Schmidt R,Geisler S and Spreckelsen C 2013 Decision support for hospital bed management using adaptable individual length of stay estimations and shared resources *BMC Medical Informatics and DecisionMaking* **13** 3.
[22] Zhu Z 2011 Impact of different discharge patterns on bed occupancy rate and bed waiting time: a simulation approach *Journal of Medical Engineering and technology.* **35** 6-7, p.338–343.
[23] Demeester P, Souffriau W, Causmaecker P D and Berghe G V 2010 A hybrid tabu search algorithm for automatically assigning patients to beds *Artic Intelligence Med* **48**. p.48-61.
[24] Brailsford S C, Viana J, Rossiter S, Channon A A and Lotery A J 2013 Hybrid simulation for health and social care: the way forward, or more trouble than it's worth? *Winter simulation conference.*
[25] Abu-Taieh E M and El Sheikh A A R 2007 Commercial simulation packages: a comparative study *International Journal of Simulation.* **8** 2.

Using Simulation to Examine the Effect of Physician Heterogeneity on the Operational Efficiency of an Overcrowded Hospital Emergency Department

Y.-H. Kuo[1] J.M.Y. Leung[2] C.A. Graham[3]
[1] Stanley Ho Big Data Decision Analytics Research Centre
The Chinese University of Hong Kong, Shatin, New Territories, Hong Kong
E-mail: yhkuo@cuhk.edu.hk
[2] Department of Systems Engineering and Engineering Management
The Chinese University of Hong Kong, Shatin, New Territories, Hong Kong
E-mail: janny@se.cuhk.edu.hk
[3] Accident and Emergency Medicine Academic Unit
The Chinese University of Hong Kong, Shatin, New Territories, Hong Kong
E-mail: cagraham@cuhk.edu.hk

Abstract. In this paper, we present a case study of modelling and analyzing the patient flow of a hospital emergency department in Hong Kong. The emergency department is facing the challenge of overcrowding and the patients there usually experience a long waiting time. Our project team was requested by a senior consultant of the emergency department to analyze the patient flow and provide a decision support tool to help improve their operations. We adopt a simulation approach to mimic their daily operations. With the simulation model, we conduct a computational study to examine the effect of physician heterogeneity on the emergency department performance. We found that physician heterogeneity has a great impact on the operational efficiency and thus should be considered when developing simulation models. Our computational results show that, with the same average of service rates among the physicians, variation in the rates can improve overcrowding situation. This suggests that emergency departments may consider having some efficient physicians to speed up the overall service rate in return for more time for patients who need extra medical care.

1. Introduction

Emergency department (ED) overcrowding is a longstanding and serious issue confronting many countries and cities around the world (e.g., [1-5]), and this phenomenon is observed to be deteriorating [6, 7]. Originally, EDs are established to provide immediate medical care to critically ill or severely injured patients, who have a very high risk of morbidity and mortality if not treated speedily and properly. Thus, timeliness and efficiency are their core attributes [8]. However, due to various causes of overcrowding, e.g., increased complexity and acuity of patients presenting to EDs, overall increase in patient volume, managed care problems and lack of beds for patients admitted to hospitals [9], EDs sometimes are not capable of delivering timely medical care to patients. Failure to provide timeous and adequate medical care to patients can be life-altering, and even life-ending: causing undue injuries and even unnecessary deaths of emergency patients. Besides, ED overcrowding can also result in other negative outcomes such as public safety at risk, prolonged pain and suffering, and long waits and dissatisfaction of patients [9]. Although both academics and practitioners have been trying to improve ED efficiency for decades, many EDs are still crowded on a daily basis. A main reason for ED overcrowding being insoluble is that most solutions to the problem are in general costly, e.g., replacing and upgrading ED facilities and expanding workforce. Most non-profit hospitals have only limited financial budgets so that these options are infeasible to them. For this reason, hospital administrators need to explore other alternatives to resolve the overcrowding problem. One way to relieve the poor situation is to enhance ED operational efficiency, for example assuring that valuable resources, such as physicians and nurses, are well-utilized.

The Prince of Wales Hospital (PWH) is one of the largest public general hospitals in Hong Kong and also the teaching hospital for the Faculty of Medicine of the Chinese University of Hong Kong. It has around 1,500 hospital beds and 4,500 staff, and serves the region of New Territories East (more than 1.5 million people). The ED of PWH, which provides 24-hour Accident and Emergency (A&E) services and has an annual attendance of 157,719 [10], is also facing the challenge of overcrowding. Due to the high quality of emergency care and the very low service charge (HKD 100, around USD 12.5, per attendance), there is often inappropriate use of the ED by non-urgent patients [11]. All of these together with the very serious understaffing situation have led to the ED overcrowding problem [12]. As a result, non-critical patients may need to experience a very long waiting time of several hours. The problems of overcrowding and long patient-waiting-times not only worsen patient satisfaction, but also may be barriers to patient access to appropriate and timely medical care.

Our project team was requested by a senior consultant at the ED of PWH to analyze the patient flow and help improve their daily operations. Our project goals are:

i. Measure and analyze patient flow and throughput of the ED.
ii. Model the daily operations and identify bottlenecks in the system.
iii. Evaluate possible changes in the processes that might enhance the ED performance.
iv. Improve patient flow and also patient experience.

Some of our previous work on this project can be found in [13].

Developing an analytical model for the entire system of an ED is very challenging due to uncertainty (e.g., stochastic patient arrivals and service durations) and the very complicated system (e.g., different categories of patients, non-stationary arrival rates, and re-entrant flows to the many "service stations"). For this reason, we adopt a simulation approach to investigate the processes of the ED and, by using the model, analyze the patient flow. This allows us easily to incorporate randomness into the system and to evaluate impacts of various changes in the system on the ED performance. From a practical standpoint, a simulation model is particularly helpful to hospital administrators and practitioners, who are not necessarily equipped with advanced mathematical knowledge, as they can effortlessly understand the model logic from the simulation animations and can examine many "what-if" scenarios, with outputs of their familiar key performance indicators (KPIs), within a user-friendly graphical interface.

Our paper is organized as follows. In the next section, we give a brief literature review on using simulation to model ED systems. In section 3, we describe our ED simulation model. In section 4, we conduct a computational study of the impact of physician heterogeneity on ED performance. In section 5, we conclude our work.

2. Brief Literature Review

Simulation has been used for about half a century (e.g., [14]) and is still one of most commonly-employed tools for analyzing healthcare delivery systems. For an overview of the applications of simulation to healthcare management, we refer the reader to [15-18].

According to [19], the number of articles related to the use of simulation in the domain of healthcare management is expanding at an incredible rate of 30 articles per day. Surprisingly, only 8% of the articles actually applied simulation to a real healthcare problem where real data was used [20]. This indicates that researchers still need a substantial effort to promote real implementations of simulation in the healthcare industry.

Here we review some of those successful applications. Rossetti et al. [21] used simulation to examine different physician work schedules and evaluate their impacts on patient throughput and resource utilization in an ED in Charlottesville, Virginia. Connelly and Bair [22] developed a simulation model for an ED in Davis, California. They used the model to compare the fast-track triage approach with a proposed acuity ratio triage (ART) approach and found that the ART approach can reduce the average treatment times for high-acuity patients but increases the average service time for low-acuity patients. Yeh and Lin [23] used simulation and genetic algorithms for nurse rostering to reduce patient waiting time in a hospital ED in Central Taiwan. Ahmed and Alkhamis [24] developed a decision support system, which integrates simulation with optimization, to determine the staffing levels of doctors, lab technicians and nurses of an ED in Kuwait. The objectives were to maximize patient throughput and to minimize patient time in the system subject to the budget constraints. The system enables the decision makers at the hospital to examine the impacts of different staffing levels on operational efficiency. Wang et al. [25] built a simulation model for an ED in Lyon. They used the model to identify process bottlenecks and evaluate how resources allocation impacts the ED performance. Abo-Hamad and Arisha [26] developed an interactive simulation-based decision support system to help improve operations of an ED of an adult teaching hospital in North Dublin. They used the tool to examine the effects of adjusting the number of staff and the physical capacity and incorporating a 'zero-tolerance' policy regarding exceeding the national 6-hour boarding time. They found that the latter has a greater impact on reducing the average length of stay of patients.

All of the above examples demonstrate that simulation is an appropriate and a very powerful tool to mimic ED systems, has a wide range of applications, and can aid hospital administrators in decision making.

3. Patient Flow and Simulation Model

In this section, we describe daily operations of the ED of PWH and our simulation model.

To provide 24-hour A&E services, the ED employs different shifts of physicians to cover the patient demand over a whole day. There are mainly three shifts: morning (M), afternoon (A), and night (N). From 07:00 to 23:00, the ED is internally divided into two divisions, the Walking division and the Non-walking division, respectively treating mobile patients and patients on a trolley or wheel-chair. However, from 23:00 to 07:00, the walking patients are directed to the non-walking division and merged with the non-walking patients to receive treatments. This policy aims to well-utilize the reduced workforce commensurate with the low arrival rates of patients in nighttime.

In the ED, there are 5 categories (levels of urgency) of patients: 1 (critical), 2 (emergency), 3 (urgent), 4 (standard) and 5 (non-urgent). In our simulation and the rest of this paper, we put category 5 patients, who are only a very small portion of the overall patients, into category 4 because they have the same flow and priority in the actual ED system. Category 1 and 2 patients have the highest priority, and category 3 patients have a higher priority over category 4 patients. Within the same category, patients are treated on a first-come, first-served basis. For critically ill and severely injured patients with life-threatening condition (i.e., category 1 and 2 patients), they are usually brought by ambulances and must be immediately sent to the resuscitation rooms and given prompt medical treatments. For the other patients, after registration, they are examined by a triage nurse and be assigned a triage category for prioritizing their levels of urgency. After triage, they stay in the waiting area until being called for consultation with a physician. After consultation, some of them may receive diagnostic tests (e.g., X-ray, blood test and CT scan), followed by a second consultation. After all the treatments are done, they either leave PWH or are admitted to the in-patient wards.

Based on the above patient flow and the ED layout, we developed a simulation model of the ED using simulation software ARENA. Our simulation model captures all the key treatment process (such as triage, consultation and diagnostic tests), the complexities of intertwining and re-entrant patient flows and time-varying and category-dependent arrival rates. The simulation model inputs include time-varying arrival rates, probability distributions for service durations, resources and physician work schedules. We presented our model to the senior consultant who initiated the project. He believed that the model was sufficient to capture all the key activities and agreed with those simulated values of KPIs. Our simulated results were also consistent with the findings in an independent research of a 5-year study of PWH ED [27]. For more details of the development of the simulation model, further validation and previous work, we refer the reader to [13].

4. Simulation Study: Impact of Physician Heterogeneity on Emergency Department Efficiency

In the literature on using simulation to model ED operations, most papers assume that all physicians are homogeneous. However, in practice, this is nearly impossible. Many medical research studies found that physician speciality has a significant impact on clinical outcomes [28, 29]. While specialists are more knowledgeable about conditions encompassed within their speciality and are likely to use more medical diagnostic tests and procedures to treat patients to improve clinical outcomes [30], generalists are more efficient in the sense that they can shorten length of stay of patients [31]. Furthermore, physician experience is also another important factor in the improvement of patient health outcomes [32]. Some recent research found that adding more junior physicians may even worsen ED overcrowding situation [33]. All of these studies suggest that physician heterogeneity would influence ED performance and should be considered when building simulation models.

We conduct a computational study using our simulation model to examine the impact of physician heterogeneity on ED performance, which is one of the concerns raised by the ED. By physician heterogeneity, we mean their efficiency and hence the service rate, which are key factors for enhancing ED effectiveness. We use the historical data of one month provided by the ED to estimate the patient arrival rates. The actual physician work schedule is used to set the shift times of physicians. Because of privacy concerns, we do not use the actual service durations of individual physicians of the ED in this experimental study. Instead, we consider three widely employed service-time distributions, which are exponential, normal and lognormal distributions, for modelling physician consultation durations. For normal and lognormal distributions, we set the standard deviation as a quarter of the mean service time. For each class of distribution, we consider three different scenarios: (i) homogeneous service rate among all physicians in the same division, (ii) heterogeneous service rates which are equal to the rate in (i) or differ from that rate by one minute, and (iii) heterogeneous service rates which are equal to the rate in (i) or differ from that rate by two minutes. The big changes in service rates, in a relative sense, would allow us to observe more clearly the impact of physician heterogeneity on ED performance. For scenarios (ii) and (iii), we set the numbers of more efficient and less efficient physicians the same for each shift and in each division. This aims to maintain the same average service rates among the physicians as in scenario (i). The consultation duration distributions for the nine instances are listed in Table 1.

Table 1. Probability distributions for consultation durations.

Instance	Walking division	Non-walking division
Exp0	Exp(5 [a, b]): M[c] (2)[d], A(2)	Exp(12): M(4), A(4), N(3)
Exp1	Exp(6): M(1), A(1) Exp(4): M(1), A(1)	Exp(13): M(2), A(2), N(1) Exp(11): M(2), A(2), N(1) Exp(12): N(1)
Exp2	Exp(7): M(1), A(1) Exp(3): M(1), A(1)	Exp(14): M(2), A(2), N(1) Exp(10): M(2), A(2), N(1) Exp(12): N(1)
Norm0	Norm(5 [a],1.25[b]): M(2), A(2)	Norm(12,3): M(4), A(4), N(3)
Norm1	Norm(6,1.5): M(1), A(1) Norm(4, 1): M(1), A(1)	Norm(13,3.25): M(2), A(2), N(1) Norm(11,2.75): M(2), A(2), N(1) Norm(12, 3): N(1)
Norm2	Norm(7,1.75): M(1), A(1) Norm(3,0.75): M(1), A(1)	Norm(14,3.5): M(2), A(2), N(1) Norm(10,2.5): M(2), A(2), N(1) Norm(12, 3): N(1)
LogNorm0	LogNorm(5 [a],1.25[b]): M(2), A(2)	LogNorm(12,3): M(4), A(4), N(3)
LogNorm1	LogNorm(6,1.5): M(1), A(1) LogNorm(4, 1): M(1), A(1)	LogNorm(13,3.25): M(2), A(2), N(1) LogNorm(11,2.75): M(2), A(2), N(1) LogNorm(12, 3): N(1)
LogNorm2	LogNorm(7,1.75): M(1), A(1) LogNorm(3,0.75): M(1), A(1)	LogNorm(14,3.5): M(2), A(2), N(1) LogNorm(10,2.5): M(2), A(2), N(1) LogNorm(12, 3): N(1)

[a] Average of the distribution.

[b] Standard deviation of the distribution.

[c] Physician shift type.

[d] Number of physicians in that shift whose consultation duration follows the corresponding distribution.

For each scenario, we ran 100 replications of simulation runs of 34 days, where each simulation replication started from an empty system. The first three days was a warm-up period and not included in our reported statistics, In other words, we simulated patient flows of 100 independent months, each starting from a non-empty system. For each instance, we generated a total number of around 1,332,000 patients (excluding those in the warm-up period) in the 100 replications. We recorded ED KPIs including the average number of patients in the ED (also known as WIP, work in process), net time for patients from registration to consultation, which we regard it as the patient waiting time (to first consultation), and physician utilization. Then we calculated their averages among the 100 replications, as reported in Table 2. Figures 1 to 4 respectively present box plots of WIP, waiting time for category 3 non-walking patients, waiting time for category 4 walking patients and physician utilization.

Table 2. Average WIP, waiting time and physician utilization.

	WIP (persons)	Waiting time (min) Cat. 3NW	Cat. 3W	Cat. 4NW	Cat. 4W	Physician Utilization
Exp0	77.97	18.03	26.56	92.03	281.32	0.93
Exp1	67.33	18.03	27.18	89.36	215.46	0.92
Exp2	51.55	18.04	26.06	89.36	117.38	0.88
Norm0	68.27	15.09	25.43	78.51	234.46	0.93
Norm1	58.38	15.15	25.31	77.91	169.85	0.91
Norm2	46.53	15.16	25.08	77.1	95.64	0.87
LogNorm0	67.12	15.03	25.14	77.23	227.04	0.93
LogNorm1	59.19	15.35	25.48	79.5	174.79	0.91
LogNorm2	46.05	14.86	25.39	74.03	94.24	0.87

From Table 2, Figures 1, 3 and 4, we observe that staff heterogeneity has a significant impact on WIP, waiting time for category 4 walking patients and physician utilization. An interesting finding is that, for all the three probability-distribution classes, the average number of patients in the ED (i.e., WIP) decreases as the variation in physician service rates increases. i.e., the ED is less crowded when such variation increases. This phenomenon also appears in the columns of average waiting time for category 4 walking patients and physician utilization. The reason for having reductions in WIP and waiting time for category 4 walking patients is that, the more efficient physicians see more patients than the less efficient physicians do. As a result, the overall service rate for consultation increases as a higher weight is placed on the service rate of more efficient physicians. This speeds up the overall consultation time and thus results in fewer patients in the system and a shorter waiting time for category 4 walking patients. The physician utilization decreases as the variation in physician service rates increases because the overall consultation time is less but the patient arrival rate remains the same. However, category 3 patients and category 4 non-walking patients do not have significant change in waiting time in this experiment as they have already had a higher priority to receive consultation. Another observation is that exponential distributions appear to have higher values of those KPIs with normal and lognormal distributions of the same averages. This is because in this set of experiments exponential distributions have higher variances than normal and lognormal distributions do. A large variance in consultation duration (i.e., a higher uncertainty) would worsen the ED performance.

The computational results show that physician heterogeneity has a significant impact on ED KPIs, in particular the patients in the ED and the waiting time for patients of low priority. Therefore, individual service rates of physicians should be taken into account in order to build more accurate simulation models for EDs. The results also suggest that, with the same average service rate among physicians, variation in physician service rates can help relieve the ED overcrowding situation: the number of patients in the ED and their waiting times can be reduced. This implies that efficient physicians can compensate for more than the inefficiency caused by the less efficient physicians. Thus, to improve operational efficiency, EDs may consider having some efficient physicians to reduce the overall consultation time but at the same time having some physicians spending more time with patients who need extra medical care. However, EDs will need to think in advance about fairness issues as the efficient physicians need to see more patients and therefore have heavier workloads. If the fairness issues are not properly handled, they will lead to low morale of physicians and may even worsen the quality of medical services provided.

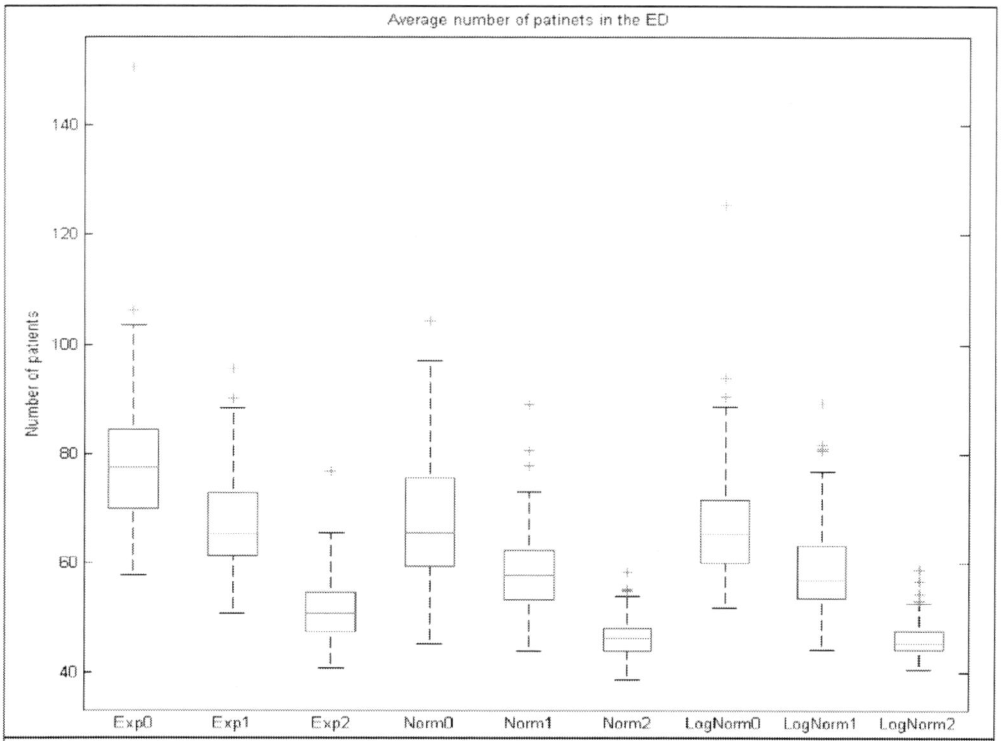

Figure 1. Average number of patients in the ED.

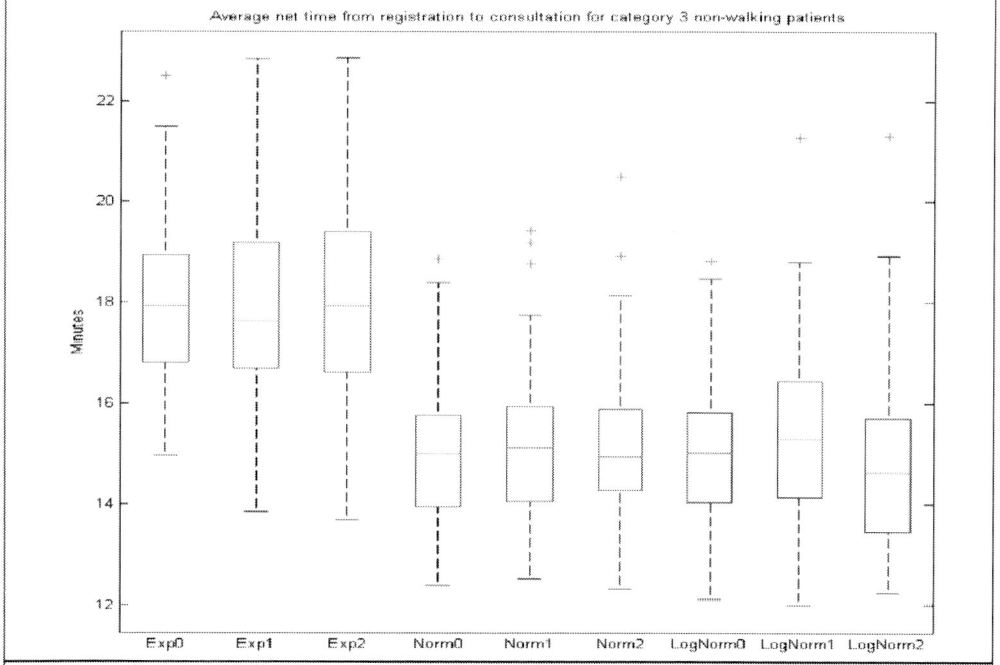

Figure 2. Average waiting time for category 3 non-walking patients.

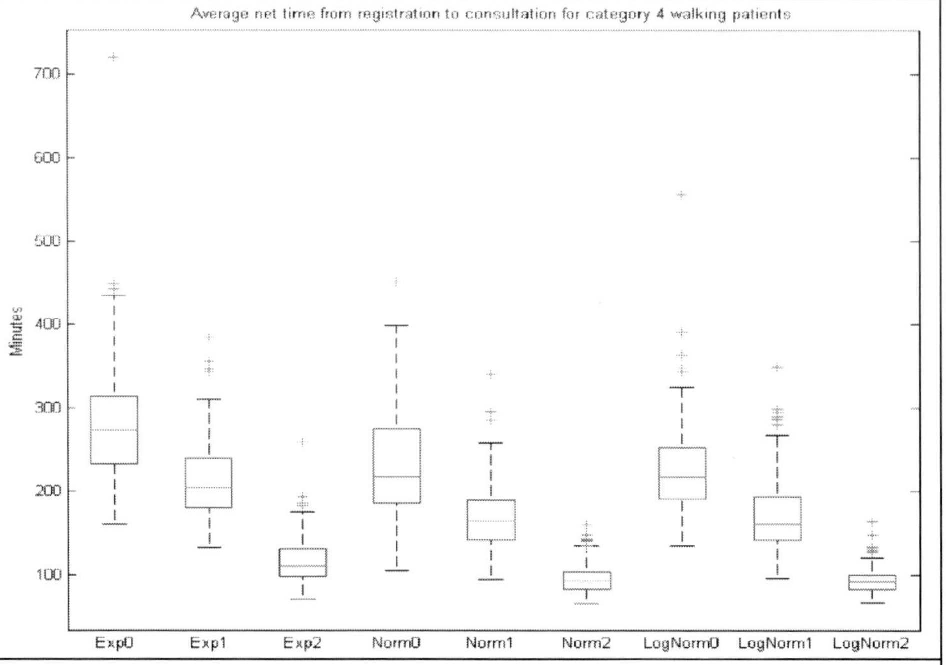

Figure 3. Average waiting time for category 4 walking patients.

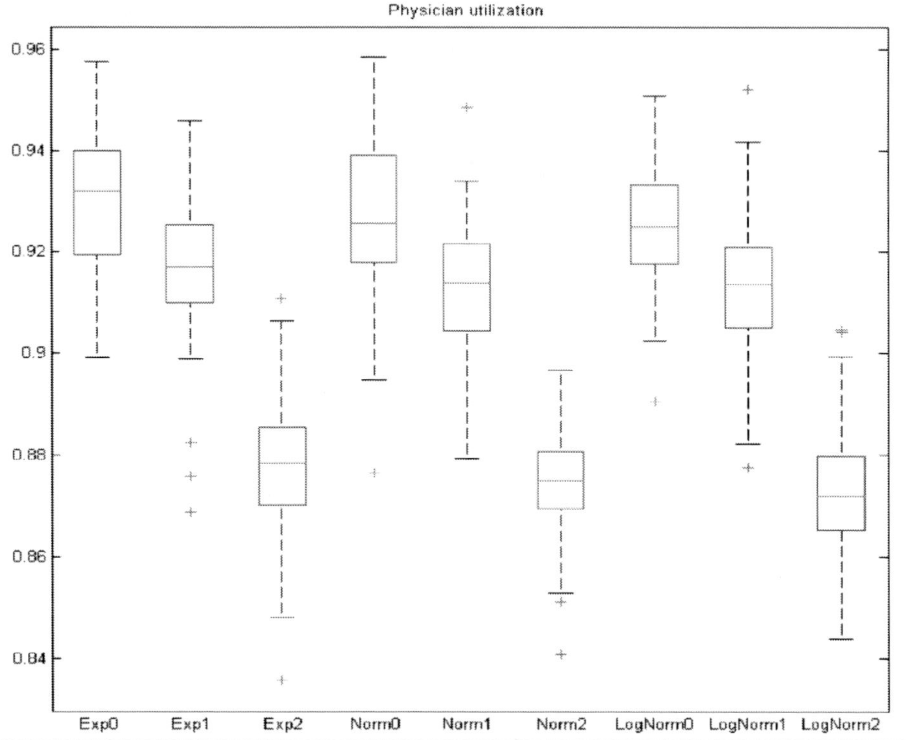

Figure 4. Average physician utilization.

5. Conclusions

Overcrowding is one of the most serious and challenging issues for EDs. Both academics and practitioners have been trying different ways to resolve this phenomenon for decades but the problem remains unsolved in many EDs. In this paper, we present a case study of using simulation to analyze patient flow of a hospital ED in Hong Kong. With our simulation model, we conduct computational experiments to examine the effect of physician heterogeneity on ED performance. We found that physician heterogeneity has a significant impact on operational efficiency. Thus, individual consultation durations of physicians are crucial when building simulation models for EDs. Computational results also show that, with the same average service rate among physicians, variation in physician service rates can improve ED overcrowding situation. This suggests that EDs may wish to have some efficient physicians who can speed up the overall consultation time in return for more time for patients who need additional care.

Acknowledgements

The research of the first author is supported by Microsoft Research Asia Collaborative Research Fund FY15-RES-THEME-049 and Macao Science and Technology Development Fund 088/2013/A3. The research of the second author is partially supported by GRF grant 414313 from the Hong Kong Research Grant Council.The authors would like to thank Mr. Stones Wong, Operations Manager of the Emergency Department of the Prince of Wales Hospital in Hong Kong, for his assistance in collecting data.

References

[1] Cowan R and Trzeciak S 2005 Clinical review: emergency department overcrowding and the potential impact on the critically ill *Crit. Care.* **9**(3) 291–95

[2] Derlet R W, Richards J R and Kravitz R L 2001 Frequent overcrowding in US emergency departments *Acad. Emerg. Med.* **8**(2) 151–5

[3] Horwitz L I, Green J, and Bradley E H 2010 US emergency department performance on wait time and length of visit *Ann. Emerg. Med.* **55**(2) 133–141

[4] Miro O, Antonio M T, Jimenez S, De Dios A, Sanchez M, Borras A, and Millà J 1999 Decreased health care quality associated with emergency department overcrowding *Eur. J. Emerg. Med.* **6**(2) 105–7

[5] Shih F Y, Huel-Ming M, Chen S C, Wang H P, Fang C C, Shyu R S, Huang G T, and Wang S M 1999 ED overcrowding in Taiwan: facts and strategies *Am. J. Emerg. Med.* **17**(2) 198–202

[6] Herring A, Wilper A, Himmelstein D U, Woolhandler S, Espinola J A, Brown D F and Camargo Jr C A 2009 Increasing length of stay among adult visits to US emergency departments, 2001 – 2005 *Acad. Emerg. Med.* **16**(7) 609–616

[7] Wilper A P, Woolhandler S, Lasser K E, McCormick D, Cutrona S L, Bor D H and Himmelstein D U 2008 Waits to see an emergency

[8] Welch S J, Asplin B R, Stone-Griffith S, Davidson S J, Augustine J, and Schuur J. 2011 Emergency department operational metrics, measures and definitions: results of the second performance measures and benchmarking summit *Ann. Emerg. Med.* **58**(1) 33–40

[9] Derlet R W and Richards J R 2000 Overcrowding in the nation's emergency departments: complex causes and disturbing effects *Ann. Emerg. Med.* **35**(1) 63–8

[10] Hospital Authority of Hong Kong Special Administrative Region 2014 *Hospital Authority Annual Report 2012 – 2013*

[11] Chung C H 2000 Emergency department misuse and administrative interventions *Hong Kong J. of Emerg. Me.* **8**(2) 151–55

[12] Rainer T 2013 Why Are Hong Kong's Emergency Departments So Understaffed? *Emergency Physicians International* **11** 26–7

[13] Kuo Y H, Rado O, Lupia B, Leung J M Y, and Graham C A 2015 Improving the Efficiency of a Hospital Emergency Department: A Simulation Study with Indirectly Imputed Service-time Distributions *Flex. Serv. Manuf. J.*

[14] Fetter R B and Thompson J D 1965 The simulation of hospital systems *Oper. Res.* **13**(5) 689–711

[15] Jun J B, Jacobson S H and Swisher J R 1999 Application of discrete-event simulation in health care clinics: a survey *J. Oper. Res. Soc.* **50**(2) 109–23

[16] Fone D, Hollinghurst S, Temple M, Round A, Lester N, Weightman A, Roberts K, Coyle E, Bevan G and Palmer S 2003 Systematic review of the use and value of computer simulation modelling in population health and health care delivery *J. Public. Health. Med.* **13**(5) 689–711

[17] Günal M M and Pidd M 2010 Discrete event simulation for performance modelling in health care: a review of the literature *J. Simulat.* **4**(1) 42–51

[18] Katsaliaki K and Mustafee N (2010). Applications of simulation within the healthcare context *J. Oper. Res. Soc.* **62**(8) 1431–51

[19] Brailsford S C, Harper P R, Patel B, and Pitt M 2009 An analysis of the academic literature on simulation and modelling in health care *J. Simulat.* **3**(3) 130–40

[20] Jahangirian M, Naseer A, Stergioulas L, Young T, Eldabi T, Brailsford S, Patel B and Harper P 2012 Simulation in health-care: lessons from other sectors *Operational Research* **12**(1) 45–55

[21] Rossetti M D, Trzcinski G F and Syverud S A 1999 Emergency department simulation and determination of optimal attending physician staffing schedules *Proc. of the 1999 Winter Simulation Conf.* pp 1532–1540

[22] Connelly L G and Bair A E 2004 Discrete Event Simulation of Emergency Department Activity: A Platform for System-level Operations Research *Acad. Emerg. Med.* **11**(11) 1177–85

[23] Yeh J Y and Lin W S 2007 Using simulation technique and genetic algorithm to improve the quality care of a hospital emergency department *Expert Syst. App.* **32**(4) 1073–83

[24] Ahmed M A and Alkhamis T M 2009 Simulation optimization for an emergency department healthcare unit in Kuwait *Eur. J. Oper. Res.* **198**(3) 936–42

[25] Wang T, Guinet A, Belaidi A and Besombes B 2009 Modelling and simulation of emergency services with ARIS and Arena. Case study: the Emergency Department of Saint Joseph and Saint Luc Hospital *Prod. Plan. Control* **20**(6) 484–495

[26] Abo-Hamad W and Arisha A 2013 Simulation-based framework to improve patient experience in an emergency department *Eur. J. Oper. Res.* **224**(1) 154–66

[27] Wai A K, Chor C M, Lee A T, Sittambunka Y, Graham C A, and Rainer T H 2009 Analysis of trends in emergency department attendances, hospital admissions and medical staffing in a Hong Kong university hospital: 5-year study *Int. J. Emerg. Med.* **2**(3) 141–148

[28] Jollis J G, DeLong E R, Peterson E D, Muhlbaier L H, Fortin D F, Califf R M and Mark D B 1996 Outcome of acute myocardial infarction according to the specialty of the admitting physician *N. Engl. J. Med.* **335**(15) 1880–87

[29] Nash I S, Corrato R R, Dlutowski M J, O'Connor J P and Nash D B1999 Generalist versus specialist care for acute myocardial infarction *Am. J. Cardiol.* **83**(5) 650–4

[30] Harrold L R, Field T S and Gurwitz J H 1999 Knowledge, patterns of care, and outcomes of care for generalists and specialists *J. Gen. Intern. Med.* **14**(8) 499–511

[31] Moloney E D, Smith D, Bennett K, O'riordan D and Silke B 2005 Impact of an acute medical admission unit on length of hospital stay, and emergency department 'wait times'. *QJM-INT. J. MED.* **98**(4) 283–289

[32] Kitahata M M, Koepsell T D, Deyo R A, Maxwell C L, Dodge W T, and Wagner E H 1996 Physicians' experience with the acquired immunodeficiency syndrome as a factor in patients' survival *N. Engl. J. Med.* **334**(11) 701–7

[33] Kawano T, Nishiyama K, and Hayashi H (2014). Adding More Junior Residents May Worsen Emergency Department Crowding *PloS One* **9**(11) e110801

AUTHOR INDEX

Institute of Physics
Dirac House, Temple Back
Bristol BS1 6BE UK

CURRAN ASSOCIATES INC
proceedings
.com